Raspberry Pi Sensors

Integrate sensors into your Raspberry Pi projects and let your powerful microcomputer interact with the physical world

Rushi Gajjar

[PACKT] open source ✳

PUBLISHING

community experience distilled

BIRMINGHAM - MUMBAI

Raspberry Pi Sensors

First published: April 2015

Production reference: 1240415

Published by Packt Publishing Ltd.
Livery Place
35 Livery Street
Birmingham B3 2PB, UK.

ISBN 978-1-78439-361-8

www.packtpub.com

Cover image by Rushi Gajjar

Credits

About the Author

Rushi Gajjar is an embedded systems hardware developer and a lifetime electronics enthusiast. He works in the field of research and development of high-speed single-board embedded computers and wireless sensor nodes for the Internet of Things. Apart from that, he studied MTech in embedded systems by being involved in research at VIT University, Vellore.

Prior to this, his extensive work as a freelancer in the domain of electronics hardware design introduced him to rapid prototyping development boards such as the Raspberry Pi. In his spare time, he loves to develop projects on Raspberry Pi that include vision, data logging, web servers, and machine learning automation systems. He loves to teach Raspberry Pi projects to school students.

His vision encompasses connecting every entity in the world to the Internet to enhance the human living experience. His hobbies include playing the tabla, photography, and travelling.

Acknowledgements

First of all, I must say thanks to my acquisition editor Richard Harvey. I never knew that I could write a book on such an amazing topic as *Raspberry Pi Sensors*. He gave me the opportunity and tremendous support and motivation before I wrote the initial chapters. So thanks for selecting me out of millions as the author of this book and inspiring me to do this.

Thanks goes also to my content editors Natasha Dsouza and Owen Roberts. You were always ready to lend me a hand wherever I was stuck. Thanks for the understanding and cooperation when I lost my pace of writing in the intermediate chapters due to tremendous work pressure. Special thanks to Natasha, who has poured hours of her time to edit the content and make it better, and supported me throughout the time I spent writing this book.

Then, thanks to my technical content developers David Alcoba, Siddharth Bhave, and Cédric Verstraeten. I sincerely thank you for investing your precious time technically reviewing this book, and providing very useful additions and valuable comments over the content, to make it more interesting for readers. By incorporating your valuable suggestions, this book has achieved a really good shape.

How can I forget Shiny Poojary and team who edited the book technically and filtered out errors in the content of the book. They totally changed the presentation of the book. I thank her for her continuous support, working untiringly to edit the book on time, and taking it to the final stage. I also thank all the employees of Packt Publishing who were directly or indirectly involved in this project, for managing everything and delivering it to the readers' hands.

Thanks to my friends, professors, and colleagues. I would like to thank all my friends, who have been part of my life, given me happiness, supported me to do this, and wished me the best before I started working on this book. Thanks to the professors at VIT University, and special thanks to Dr. Arun Manoharan for giving me a small but very helpful insight into being an author. My colleagues at Leaf Technologies always took updates from me about the progress of this book and encouraged me to include strong content.

Above all, I would sincerely like to thank my parents for asking me every day about this book's progress and showing keen interest in seeing it take shape, in spite of all the time for which it kept me away from them.

About the Reviewers

David Alcoba, for many years, considered himself a software engineer who liked to play with electronics in his spare time. While being responsible for designing and building highly secure distributed applications for the industry, he also decided to start gaining more and more knowledge of digital fabrication tools every day. And it was then that he realized he had just discovered a world where all of his different interests could be merged into a single project.

Based on this idea, he helped create Vailets Hacklab in 2014, a local community in Barcelona that aims to hack the current educational system so that kids might be co-creators of their future through technology, instead of being just its consumers.

Following the spirit of this initiative, David decided to cofound Makerkids Barcelona, a small start-up focused on providing professional services for schools and organizations to engage kids with the new maker movement and follow the STEAM (science, technology, engineering, art, and mathematics) educative principles.

Nowadays, David feels that he is not an engineer anymore but a maker.

Siddharth Bhave is a big data researcher at the Center for Data Science at the University of Washington. With a background in electronics and embedded systems, he is interested in the distributed systems aspect of recent big data technologies such as Hadoop and Spark. Siddharth implements and analyzes various machine learning algorithms on Xeon servers. Characterizing their behavior and studying the scalability of algorithms is something that he picked up during his internship at Intel. During his MS degree in computer science, he worked on developing a piece of middleware to work with real-time sensor data fed to a cluster of Raspberry Pi nodes. He wants to translate his work to scale and expand the concept of Internet of Things.

I would like to thank my family and friends, who always believe in me, and all my teachers and professors, who always inspire me.

Cédric Verstraeten holds an MSc in engineering and is primarily active in the C++ community. He works as a software engineer and is a huge open source enthusiast. He spends most of his time on side projects that can automate and simplify people's lives. He's the organizer of the Raspberry Pi Belgium meet-up.

I would like to thank Packt Publishing for allowing me to be a reviewer for this book. I really think their books can give people an in-depth overview of a particular topic.

www.PacktPub.com

Support files, eBooks, discount offers, and more

For support files and downloads related to your book, please visit www.PacktPub.com.

Did you know that Packt offers eBook versions of every book published, with PDF and ePub files available? You can upgrade to the eBook version at www.PacktPub.com and as a print book customer, you are entitled to a discount on the eBook copy. Get in touch with us at service@packtpub.com for more details.

At www.PacktPub.com, you can also read a collection of free technical articles, sign up for a range of free newsletters and receive exclusive discounts and offers on Packt books and eBooks.

https://www2.packtpub.com/books/subscription/packtlib

Do you need instant solutions to your IT questions? PacktLib is Packt's online digital book library. Here, you can search, access, and read Packt's entire library of books.

Why subscribe?

- Fully searchable across every book published by Packt
- Copy and paste, print, and bookmark content
- On demand and accessible via a web browser

Free access for Packt account holders

If you have an account with Packt at www.PacktPub.com, you can use this to access PacktLib today and view 9 entirely free books. Simply use your login credentials for immediate access.

Table of Contents

Preface

Raspberry Pi is a single-board, credit-card-sized, computer packed with many opportunities to explore and invent. It is really amazing to see kids start coding Python from scratch, and build a bird box that streams live video on the Internet to check whether a bird has got its meal. I remember that when I was a kid, I used to play with Lego toys attached to DC motors and batteries, which was engaging. At that time, I could not imagine the logic that went into coding, and did not get any chance to code my projects and control the movement of those Lego blocks using a mobile phone. But I am lucky enough now to get an opportunity to explain such projects and provide a launchpad for young creators who really have a passion to create something and change the world around us.

The world is moving towards a new era. Technology is revolutionizing daily needs and habits and making them available on a simple interface, which gave me motivation to write a book on Raspberry Pi sensors. It's a world of creativity, and the I believe that creativity comes when you start understanding and appreciating the fundamentals and start applying logic to it. A lot of information and projects on Raspberry Pi are floating on various webpages, and one wishes to achieve as much as he/she can. I feel that the information on webpages is often observed as incomplete. It gives us a quick start to build projects but does not explain what is behind them.

It is known that without actually diving too deep into electronic devices and communication protocols, you can start coding on Raspberry Pi and craft amazing projects. I have colleagues around me who often need to code and wire the sensors on the Raspberry Pi platform for their experiments. They can develop Python code on artificial neural networks in a short span of time, but when it comes to wiring something, they look around. I believe that a basic understanding of electronics is a plus for such prodigies out there, who want to develop code on such platforms. In the opposite scenario, hardware developers can wire sensors, ensures proper voltage levels on device pins, but when it is time to code, they need help.

The most interesting thing that I find with the Raspberry Pi is that I can still play with the hardware components and soldering iron, and code my hardware to make it live. This book provides a kick start for such creators, who really want to know how things work together, and want a direction for starting projects on sensor interfacing and the Internet of Things with Raspberry Pi. There is tremendous growth in technology when we look towards the connected array of everything around us.

Internet of Things opens up a new world for collecting data and analyzing it for better user experience. A lot of data from the array of sensors has been generated from several different sensor nodes. In this context, the Raspberry Pi provides us with the opportunity to start with simple projects, such as uploading data to the Internet from a developed sensor station, as described in the chapters of this book. This will be your first step to building an Internet of Things project. Another interesting thing is that with the rise of Raspberry Pi 2 model B, developers have got enough processing power to perform computation-intensive algorithms on the Raspberry Pi. Therefore, image processing has been included in one of the chapters. It would have been very difficult to try to explain image processing to beginners, but I have at least tried to offer a simple start for readers so that they begin image processing on their own.

This book explores five different projects, any of which can be a startup for different ways of building electronics projects. The approach I have followed while preparing the projects is quite interesting. This is the methodology I often follow to develop complex hardware designs. Although I do not rely on breadboards (as I am more into high-speed circuit designs), small project prototypes, some of which are covered in this book, can easily be wired on breadboards. The first approach should be to purchase the best hardware components (preferably through hole for breadboard testing), on which you can rely when the code is not working or not giving the proper results. Prepare a block diagram and consider each issue that may occur during hardware and firmware design. Second, read datasheets of components used and ensure every single entity meets the design requirements. Thirdly, wire the components to the breadboard and check it thoroughly. Finally, when the hardware is built robustly, write the code (or firmware), and rewrite it to make it more perfect. Remove the unnecessary variables and unreachable code or loops, handle interrupts, define the sleep time and watchdogs of a processor, and manage proper memory segments to avoid crashes. However, this book has followed mostly simple code that does not go that deep into managing embedded programming. Installing useful coding libraries on the Raspberry Pi takes care of the faults often created by a programmer. Just call a function and it does all the embedded calls in the background. Thanks to the developers of the Raspberry Pi libraries, with which we build more robust code (whether knowingly or unknowingly). When you prepare a sample of code, I advise you to break it down into pieces.

You may face some difficulty when building the project for getting data from temperature, humidity and light sensors. First, get it done for temperature and humidity, and then code for the light sensor. Whenever both pieces of code give you the desired values, recode them. Then you can combine them by managing the function calls.

When writing the chapters, I have followed a common theme across the book: first the setup, then the purpose of the project, and finally describing the hardware with complete details. In some of the chapters, the software has been divided into components, and then they have been merged so as to avoid monotony for you. I apologize to you for being lengthy in the theory portion in some parts of this book, but I am sure that you will love to read and learn a lot from it, and you can get the most out of it.

Any questions, improvements, and suggestions are welcome, and should probably take place in the GitHub issues for the book at `https://github.com/rushigajjar/raspberrypisensors` so that everybody can take part. Besides that, anybody can contact me on LinkedIn at `https://in.linkedin.com/in/rushigajjar`, and send messages regarding their interesting projects and startups. I would really love to hear about it. Or you can send tweets by sharing temperature and luminance values at `@rushigajjar` once you get your air conditioner and lights automated!

What this book covers

Chapter 1, *Meeting Your Buddy – the Raspberry Pi*, gives an introduction to all the models of Raspberry Pi available in the market, including the recent Raspberry Pi 2 model B. A method of installing the operating system and interesting ways to share the Internet with the Raspberry Pi are discussed. Then we perform some hands-on coding in Linux terminal, Linux shell scripting, Python, and C on the Raspberry Pi.

Chapter 2, *Meeting the World of Electronics*, explains the fundamentals of electronics and communication protocols by which electronic devices communicate. Experiments with GPIO are more interesting to perform with the shell, Python, and C languages.

Chapter 3, *Measuring Distance Using Ultrasonic Sensors*, shows you how to set up an ultrasonic sensor with the Raspberry Pi and learn to take care of different voltage levels across the devices. We prepare a code to get our ultrasonic sensor running, and develop an aid for a visually impaired person with an obstacle avoidance system.

Chapter 4, *Monitoring the Atmosphere Using Sensors*, develops your skills in choosing a sensor from many that are available in the market. We then implement the hardware and software required for temperature, humidity, and light sensors to automate our home appliances.

Chapter 5, *Using an ADC to Interface any Analog Sensor with the Raspberry Pi*, explains interfacing of analog-to-digital convertors with an array of sensors. We build a sensor station for the Raspberry Pi using serial protocols to use the generic software function built, to get data from any sensor interfaced with it. Finally, the data can be stored in a log file for analysis.

Chapter 6, *Uploading Data Online – Spreadsheets, Mobile, and E-mails*, takes a dive into the Internet of Things and sensor nodes. With the help of the sensor station developed in the previous chapter, we upload the data to online spreadsheets and observe a real-time graph. We also get emergency e-mails on our e-mail platforms. Once you get your project done, you can send your sensor values to rushi. raspberrypisensors@gmail.com.

Chapter 7, *Creating an Image Sensor Using a Camera and OpenCV*, covers the basics of image processing and how an installation of the OpenCV library can be performed successfully. Using a camera, we will develop an IP camera to install in your backyard to observe live streaming of activities. Further, we will build a piece of motion detection code in OpenCV to detect human movement in a particular area and capture an image for an immediate alert.

Appendix, *Shopping List*, includes the list of the hardware components that need to be purchased in order to perform the hands-on tasks described in the book. From chapter 2 onwards, these components will be required to test our codes. You can directly take this list to the electronics distributors near you and come home with a filled shopping bag!

What you need for this book

There are no special demands for implementing the projects of this book on the Raspberry Pi, except the essential hardware components! You can use a personal computer with Linux, Windows, or Mac OS X to connect the Raspberry Pi to it. Any of the Raspberry Pi models available in the market (Raspberry Pi 1 models A+, B, or B+, or Raspberry Pi 2 model B) should be okay for performing experiments on the code. You can refer to the *Appendix*, *Shopping List*, for the requirement of hardware before getting into *Chapter 3*, *Measuring Distance Using Ultrasonic Sensors*, and the later chapters.

Who this book is for

This book is perfect for hardware enthusiasts who want to create a variety of projects using the Raspberry Pi. It is for people who have prior programming experience, especially in Linux, C, and Python, but it's not limited to them. Those who do not have programing knowledge can get the essentials from the book and start developing the projects instantly. In any case, this book will get you ready with all the latest electronics concepts that are required for hardware programing with the Raspberry Pi.

Conventions

In this book, you will find a number of styles of text that distinguish between different kinds of information. Here are some examples of these styles, and an explanation of their meaning.

Code words in text, database table names, folder names, filenames, file extensions, pathnames, dummy URLs, user input, and Twitter handles are shown as follows: "There are two functions in the spi-dev library for sending the data to SPI slave devices."

A block of code is set as follows:

```
import RPi.GPIO as GPIO
from time import sleep
GPIO.setmode(GPIO.BCM)
GPIO.setup(17, GPIO.OUT)
Print "All Set in Python! Let's Blink"
```

Any command-line input or output is written as follows:

```
sudo nano /etc/modprobe.d/raspi-blacklist.conf
```

New terms and **important words** are shown in bold. Words that you see on the screen, in menus or dialog boxes for example, appear in the text like this: "Select the **Enable Camera** option in the configuration settings using the keyboard."

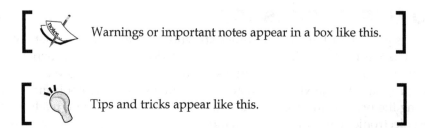

Warnings or important notes appear in a box like this.

Tips and tricks appear like this.

Reader feedback

Feedback from our readers is always welcome. Let us know what you think about this book—what you liked or may have disliked. Reader feedback is important for us to develop titles that you really get the most out of.

To send us general feedback, simply send an e-mail to feedback@packtpub.com, and mention the book title via the subject of your message.

If there is a topic that you have expertise in and you are interested in either writing or contributing to a book, see our author guide on www.packtpub.com/authors.

Customer support

Now that you are the proud owner of a Packt book, we have a number of things to help you to get the most from your purchase.

Errata

Although we have taken every care to ensure the accuracy of our content, mistakes do happen. If you find a mistake in one of our books—maybe a mistake in the text or the code—we would be grateful if you would report this to us. By doing so, you can save other readers from frustration and help us improve subsequent versions of this book. If you find any errata, please report them by visiting http://www.packtpub.com/submit-errata, selecting your book, clicking on the **errata submission form** link, and entering the details of your errata. Once your errata are verified, your submission will be accepted and the errata will be uploaded on our website, or added to any list of existing errata, under the Errata section of that title. Any existing errata can be viewed by selecting your title from http://www.packtpub.com/support.

Piracy

Piracy of copyright material on the Internet is an ongoing problem across all media. At Packt, we take the protection of our copyright and licenses very seriously. If you come across any illegal copies of our works, in any form, on the Internet, please provide us with the location address or website name immediately so that we can pursue a remedy.

Please contact us at copyright@packtpub.com with a link to the suspected pirated material.

We appreciate your help in protecting our authors, and our ability to bring you valuable content.

Questions

You can contact us at questions@packtpub.com if you are having a problem with any aspect of the book, and we will do our best to address it.

1
Meeting Your Buddy – the Raspberry Pi

The world is being automated, with huge chunks of data being produced and processed for the analytics, controlling, and connecting purposes. The **Raspberry Pi** board can provide a vast range of automation and data processing if used vigorously. This tiny board provides ample functionalities and opportunities to change the world around us. This chapter is the first step towards doing that.

If you are first-time Linux user or are new to programming, it may seem difficult to understand many commands and codes, but the motivation to change the world is likely to be enough to start building the projects. This chapter provides an easy guide to start using the Raspberry Pi board and the total build-up for the users to interface the sensors. It will make the Raspberry Pi your best buddy. It is important to know the steps included in this chapter to rapidly build the projects. This chapter covers:

- A basic understanding of the Raspberry Pi board and its useful connectors
- Steps to install the operating system for the first time
- Unique methods to share the Internet connection with the Raspberry Pi
- Linux basics and useful shell commands
- Installing important libraries
- Introduction to compiling and executing C and Python programs
- Practice problem statements for Shell, C, and Python

A glance at the Raspberry Pi board

Before we get started, it's time to reintroduce our good friend, the Raspberry Pi. Kudos to the designers of the board, who have packed everything we need to accomplish our projects in a credit-card-sized printed circuit, also called a credit-card-sized single-board computer. There are two versions of Raspberry Pi: Raspberry Pi 1 and Raspberry Pi 2. Due to earlier developments, the Raspberry Pi 1 family consists of model A, model A+, model B, and model B+. The recently launched model is Raspberry Pi 2, the new addition to the model B category. Nowadays, the most widely used Raspberry Pi is model B+, which is also called the original Raspberry Pi board in the Raspberry Pi 1 family. The predecessor of the Raspberry Pi models B and B+ was model A, which is not widely used in the hobbyist space compared to other models such as A+ and B+. If you are not aware of the specifications of these boards, take a look at the complete comparison in the following table, which contains the comparable parameters of the current models of Raspberry Pi 1 and 2. Then you can choose a board that you want.

Features	Raspberry Pi 1			Raspberry Pi 2
Models	B	A+	B+	B
Processor	BCM2835			BCM2836
Processor cores	Single			Quad
Speed	700 MHz			900 MHz
RAM	512 MB	256 MB	512 MB	1024 MB
GPU	VideoCore IV			
Pin header	26 pin	40 pin	40 pin	40 pin
Audio and video ports	RCA, HDMI port	3.5 mm jack, merged audio-composite video and HDMI port		
Ethernet port	Yes	No	Yes	Yes
USB ports	2	1	4	4
Power	Micro USB port			
Digital interfaces	CSI (camera), DSI (display) ribbon cable connectors			
Memory card	SD	MicroSD		

Raspberry Pi 1 has a Broadcom BCM2835 processor with a 256 MB or 512 MB RAM on top of it. The processors and RAM are integrated as **Package on Package** (POP). On the other hand, Raspberry Pi 2 has a Broadcom BCM2836 processor, which comes with a 1024 MB RAM interfaced beneath the board. Raspberry Pi 1 model A+ is still loved by minimal RasPi users who need low-powered performance when they are running on batteries. There's lots of good stuff here: RAM states the temporary memory available to run the current processes and applications. Multimedia processing ensures smooth graphical processing to run high-resolution videos through HDMI and video-extensive applications on Raspberry Pi.

 We have already decided to make the Raspberry Pi our friend, and like all our friends, it requires a unique and cool name so that we can call it easily when we need it. I would like to call it a RasPi, so throughout the book, whenever you are referring to the name RasPi, it's your buddy, Raspberry Pi.

Your new friend has all the capabilities that your computer has. The RasPi can be used to understand how a computer works, to learn programming, for word processing, and for gaming. Here are the small and shiny hacks that we can do with RasPi:

- Do you want to watch your favorite high-definition movie just by connecting a display to it? You can do this.

- Do you want to use RasPi as a web server, where you can run your websites? Not a problem with RasPi.

- Do you have a vacation and want to play video games, such as Minecraft? You can try using RasPi.

- Do you want to use it as your point-and-shoot digital camera while you are going to visit a zoo this weekend? Easy!

- You can even make your own robot or quadcopter using the RasPi. Wow!

All of these features come in such a small piece of board.

 Does this make you excited? Obviously, yes! There are such numerous applications that we can build with the RasPi, but they are out of the scope of this book.

Because of its ability to interact with the outside world, the major applications developed using RasPi include recognizing the surrounding parameters using sensors and converting them into useful data to analyze and control the appliances that we are going to experiment in the upcoming chapters.

I assume that you have the RasPi (model B or B+) in your hand, and you might be wondering what are the different connectors and electronic elements on the board. Rather than introducing the jargon of specifications, I will introduce what we need to make our projects. Take a look at the different connectors in the following diagrams. The nomenclature presented here will be used throughout the book.

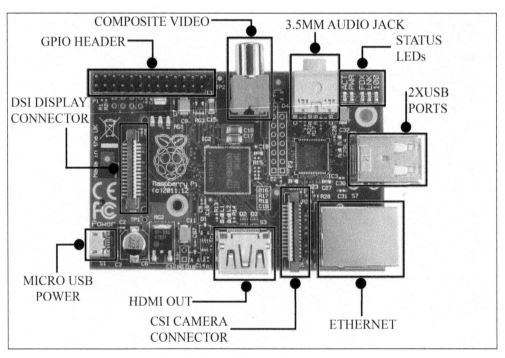

The Raspberry Pi 1 model B connectors

The Raspberry Pi 1 model B+ and Raspberry Pi 2 model B look identical to each other, and the difference is only in performance.

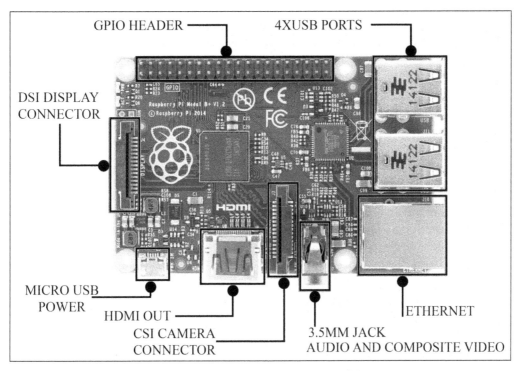

GPIO HEADER ——— 4XUSB PORTS ———

DSI DISPLAY
CONNECTOR

MICRO USB
POWER
HDMI OUT ———
CSI CAMERA
CONNECTOR

ETHERNET

3.5MM JACK
AUDIO AND COMPOSITE VIDEO

The Raspberry Pi 1 model B+ and Raspberry Pi 2 model B connectors

Due to a more powerful processor and an upgraded RAM, Raspberry Pi 2 Model B improves on performance by six times. In a clockwise direction in the diagram of the RasPi 1 model B, the short description of the important connectors is as follows:

- **GPIO header**: GPIO stands for **General Purpose Input Output**, which has been brought out to pin connectors present on the board. The processor on board (BCM283x, which is the brain of a RasPi) has a facility to provide a specific functionality during the runtime of your own program. We are going to use them a lot in the upcoming chapters. The great thing with this is that you can assign a specific task to the specific GPIO in your program, and while your program executes, it goes to logic low or high (triggers to off state and on state) accordingly. We can read values from any other peripherals, such as sensors, and compute the received values in your own programs. Apart from reading the values, we can show the result of the program by connecting LEDs or embedded LCD displays to the board. Depending on the decision taken in the code, we can drive a motor connected on GPIO through a motor driver circuit. This feature on RasPi makes a huge difference compared to the normal computing board by giving developers the freedom of crafting the creation.

For example, in one of your applications, if the temperature falls below 20 degree Celsius then the thermostat connected to your RasPi gets the signal through the specific GPIO assigned and it starts heating. GPIOs typically work in logic high (1, ON) and logic low (0, OFF), and this will work the way you program it!

- **RCA video out**: This is the most widely used and one of the oldest connectors that both old and new televisions or displays use. It carries the video signal, which is the type output on the RasPi. The RCA connector or composite video signal is merged with a 3.5 mm audio jack on RasPi 1 model A+ and model B+ and RasPi 2 model B.

- **3.5 mm audio out jack**: If you are not using the HDMI connection (which will be described soon), the audio can be played through speakers or headphones using a standard 3.5 mm jack. In RasPi 1 model B+ or RasPi 2 model B, audio jack being the combination of composite and audio has all the functionalities of composite video and audio out.

- **USB**: This is the most common connector, widely used in the modern computers, and hence called the **Universal Serial Bus**. You can connect your flash drives, keyboard, Wi-Fi dongles, and mouse to play around with the RasPi. You can also connect the externally powered USB hub with RasPi to connect more USB-based peripherals on it.

- **Ethernet**: This is one of the most important connections to have a remote login on RasPi and to provide wired internet connection. In the next sections of this chapter, we will be using it widely. We cannot always connect RasPi to the dedicated display, so we use the remote login, and we see the entire desktop or **Command-line Interface (CLI)** of RasPi on our computer screen.

- **CSI camera connector**: The RasPi board does not come with camera module integrated, but a separately bought camera module can be interfaced using the CSI connector via a 15 cm flex cable. A longer flex cable will lead to bad quality of images. The 5-megapixel Raspberry Pi camera module can be used to record high-definition videos as well as still photographs. It's easy to use for beginners, but has plenty to offer advanced users if you're looking to expand your knowledge. This camera module provides improved performance over a USB-connected camera.

- **HDMI connector**: The **High-definition Multimedia Interface (HDMI)** is a compact audio/video interface used to transfer uncompressed media data. You can connect your modern HDTV to watch full high definition (FHD/HD) videos through the RasPi. If you plug in the HDMI connector, there is no need to connect the speakers to the audio jack, and if you want to get sounds on both HDMI and the 3.5 mm jack, then you'll have to play with and edit the internal files of Linux.

- **Micro USB power**: You survive on food, don't you? Well, so does the RasPi (kind of). It needs power supply to operate. The device can be powered by a 5V input voltage, and the current ratings solely depend upon what you have hooked up with RasPi. Have you seen any power button on RasPi? In fact, the RasPi module does not have the power on button. Therefore, just plugging the micro USB power adapter will boot the RasPi.

The maximum current the Raspberry Pi models A and B can use is 1 ampere, so if you need to connect a USB device that will take the power requirements of the Raspberry Pi above 1 ampere, then you must connect it to an externally powered USB hub. For example, a USB hard disk will need an ample amount of current to operate, which RasPi cannot deliver through the USB port. Alternatively, the maximum power model B+ can use is 2 amperes before needing to connect devices to an externally powered USB hub. There are power banks and batteries available for connecting to the RasPi if you are designing a remotely operated car or a quadcopter. If you are not sure how much power the USB device is going to take, buy an externally powered USB hub. Do not go above 2.4 amperes in any case, because this will destroy your RasPi if peripheral current demand is high—it'll be dead!

- **SD card slot**: The SD card is important because it is where the RasPi keeps its operating system. It is also where you will store your documents, programs, and pictures. It is the secondary and a necessary memory part for the RasPi, the on-board RAM being the primary. Model B requires the standard-sized SD card (the big one!), whereas model B+ requires the microSD card.

It is suggested to purchase the microSD card with the SD card adaptor so that if you switch over from RasPi 1 model B to B+ or RasPi 2 model B, you can retain the same operating systems and your programs. Additionally, after installing the libraries and setups, the OS crash can be painful. To avoid this, the periodical backup of the entire OS should be taken, and this can be used to install the OS on a new SD card again. The microSD card can easily be contained in an SD adaptor to convert it into a normal SD card, with no performance losses.

- **Display connector**: Last but not least, the display connector is used to connect a 7-inch finger-touch LCD display to the board for your embedded product development. But usually, the RCA and HDMI are enough. If your application requires this, then you will need to use it.

Setting up for the first time

Following your recent purchase of the RasPi, you now have to make it ready to work for you. In this section, we are going to install an operating system on which our RasPi will run. The most popular, stable, and widely used operating system for RasPi is the Wheezy-Raspbian.

The Raspbian runs on an open platform, Debian Linux. Why are we going to use Raspbian instead of directly using Linux and other flavors of Linux? Simply because Raspbian has all device drivers written for the RasPi. In brief, the device driver is a program that gives the details of the hardware to the running operating system and supports the user interface to take/give commands from/to the hardware. As our RasPi has different hardware than a personal computer or general-purpose computer, the modified operating system is needed to completely use all resources on the hardware. It is best for those who want to follow the default standards.

You can download the Wheezy-Raspbian install file (known as an image file) from the Raspberry Pi foundation's website at `http://downloads.raspberrypi.org/raspbian_latest`. If you want to install the GUI-free flavor of Linux (Direct CLI), you can try Arch Linux. It can be downloaded from the ArchLinuxArm web page at `http://archlinuxarm.org/os/ArchLinuxARM-rpi-latest.tar.gz`.

Installing the operating system

We need some essential components to successfully start up the RasPi. Note that this subsection is for those users who are using the RasPi for the first time and have not purchased the preloaded SD card. If you have installed the operating system on your SD card, then you can skip this section, or just look at the procedure to help your friend who has just brought a new RasPi.

Purchasing your SD card

When you buy a RasPi module, it may or may not be sold with an SD card. If your RasPi did not come with an SD card, then the minimum size you should get is 4 GB. RasPi doesn't have any on-board memory, so purchasing an SD card is the only way to get data storage as well as an operating system. I have an 8 GB SD card for my RasPi, and it works well and is sufficient for doing almost all projects.

 For server applications, if you require more space, then an SD card with higher space will be useful.

Downloading the required software

Once you have an SD card, you are ready for this step. These steps are broadly explained in my personal blog too, which you can visit at `http://rushigajjar.blogspot.in/2014/03/setting-up-raspberry-pi-for-first-time.html`.

Let's see the procedure for different operating systems.

Windows

These things need to be kept in mind while working on Windows:

- **Format the SD card**: Use the SD Card Association's tool (`www.sdcard.org/downloads`) to format/wipe your old data from the SD card. Use of the SD formatter provides optimal performance for your memory cards compared to the generic formatting facilities provided by your computer's operating system.

- **Write the image of the OS on an SD card**: The downloaded Raspbian OS will be written from a raw disk image to a removable device using the popular and free tool, Win 32 Disk Imager (`sourceforge.net/projects/win32diskimager/`). Follow the steps on the screen, and you can easily write an image of Raspbian on your SD card. Once you are done with this process, jump to the next section.

Mac OS X

The following things need to be kept in mind while working on Mac OS X:

- **Download software to write the image of the OS on an SD card**: There are multiple tools available for Mac OS X users such as ApplePi-Baker, PiWriter, and Pi Filler. PiWriter is a CLI-based tool, whereas ApplePi-Baker and Pi Filler are GUI-based tools. Pi Filler is recommended to be used because it is simpler and faster than other tools. It can be downloaded from `http://ivanx.com/raspberrypi/files/PiFiller.zip`.

- **Write the image on the SD card**: Insert an SD card, locate or choose the downloaded image in the tool, and erase the SD card in the automatically popped-up menu. Select the **continue** button to write the image on the SD card.

Linux

These things need to be kept in mind while working on Linux:

- **Unzip the downloaded image**: If you are using a GUI-based Linux OS on your desktop computer, open the `download` folder and unzip the OS by right-clicking on **Extract Here**.

- **Download the software to write the image of the OS on an SD card**: You can install image writer from the Ubuntu Software Center. Open **Software Center** and search for **ImageWriter**. Insert the SD card in your desktop and follow the GUI (click on the **Write to Device** button) to locate and write the image on your SD card.

Expanding the root filesystem

Now, it's time to start the RasPi for the first time. After the process of writing an OS on the SD card, insert the SD card into the slot available on the RasPi. Connect the display (or TV) through an RCA or HDMI connection, and power up the RasPi by connecting a power supply to the micro USB connector. You will be able to see the configuration screen. Directly select the **expand_rootfs** option (from the keyboard connected through the USB of the RasPi) and wait for some time to complete the background process. When you write the image on the SD card, everything is in a compact bundle, where it needs to be expanded for the RasPi's complete operation. Expanding the root filesystem (expand_rootfs) resizes the partitions in the SD card and allows us to utilize the memory space in it. By this point, you should restart your RasPi.

Before powering up the RasPi using the micro USB cable, the SD card must be inserted and the HDMI cable must be attached. RasPi reads the signals coming through HDMI to check the display connection. If the display is not available at the first point of startup, it disables the HDMI interface and streaming to optimize performance.

Logging in to the RasPi

When the RasPi restarts, you will be able to see many pieces of code running on the screen. Don't panic! It's a normal process that goes on in the RasPi. When it becomes stable, it will prompt you to enter the username and password. By default, the username and password are pi and raspberry respectively.

Note that while you are writing your password, you will not be able to see anything coming on the screen; don't worry. Welcome to the Linux world!

Once you've entered the correct password, you will be able to see the CLI with pi@raspberrypi~$ on your screen, monitor, or TV, which is now ready to take commands from you. Aye, aye, captain!

Opening the desktop

Enter the `startx` command as `pi@raspberrypi~$startx` and press *Enter*. Now you have a white screen with the Raspberry Pi logo and a GUI that looks similar to our personal desktop computers. Take a bite!

So, we saw an easy and compact guide for setting up the RasPi for the first time. We'll now add more functionality to our RasPi by providing an Internet connection for it. A computer is incomplete without an Internet connection, and so is our RasPi. This is something we need to solve, especially as we can directly download useful libraries and applications on the RasPi through the internet.

Connecting the Raspberry Pi to the Internet

Once you have finished setting up the RasPi, it's time to connect it to the Internet. Basically, there are two very common ways of connecting the RasPi to Internet: the first (and easiest way) is via Wi-Fi connection using a Wi-Fi dongle, or transceiver; the second is somewhat tricky but it's the most economical and practical way to utilize everything you have and without an additional Wi-Fi dongle. This will require a laptop/desktop computer (a PC) and an Ethernet cable. We will see how to follow each of the ways.

Internet connection through Wi-Fi dongle

You can purchase the dongle for the RasPi from any leading online store or an electronics store near you. It ranges from 10 to 20 USD at the time of writing this book. While in the process of purchasing, read about the power requirements of the dongle. You can purchase the miniature Wi-Fi dongle available on Adafruit, or a dongle from brand names: PiFi or Edimax. If you are thinking of giving a try to Ethernet Internet connection sharing, then this section can be skipped. The basic steps for enabling the Wi-Fi network connection are introduced here.

It is advisable to use either of the methods at once. If you choose to use the Wi-Fi dongle, then you can skip the section of Ethernet sharing and vice versa.

In the CLI of the RasPi, enter the following command to note down the gateway and netmask of the Ethernet connection so that you can add a static IP defined in the interface file in the upcoming steps:

```
netstat -nr
```

We have to perform the following steps to enable the Wi-Fi network connection:

1. Go to the network interfaces file of the RasPi by entering the `sudo nano / etc/network/interfaces` command in the CLI. Note that you will get acquainted with these commands in the upcoming sections. Once you enter the command, the text you need to change is this:

```
auto lo
iface lo inet loopback
iface eth0 inet static
address 169.254.0.2
netmask 255.255.0.0
broadcast 169.254.0.255
gateway 0.0.0.0

allow-hotplug wlan0
iface default inet dhcp
        wpa-ssid "ssid_name"
        wpa-psk "password"
```

 Do not forget to put `ssid` (your network name) and your password (you know it!) in the quotes.

2. After editing the file, press *Ctrl* + *X* and press *Y* to confirm the edit made by you.

Shut down the RasPi by entering the `sudo poweroff` command. You then need to connect the Wi-Fi dongle and turn it on again. While it is booting up, it finds the Wi-Fi adaptor connected to it. Pretty simple, isn't it!

Internet connection through Ethernet from a PC

All of the preceding steps require a dedicated display, mouse, keyboard, and all other cables to get the view of the working RasPi. For regular uses, this is somewhat bulky to carry all of these along with your RasPi. At this stage, I assume that you have already installed Raspbian OS.

For this method, you just need your laptop/desktop (it already has an inbuilt Wi-Fi module, which is why we don't purchase an additional Wi-Fi dongle for RasPi until we have a special requirement), the Windows operating system, and an Internet connection. You do not need any add-on displays, keyboard, mouse, or Wi-Fi dongles connected with the RasPi.

So all we need is an Ethernet cable, a power supply to RasPi, the SD card with Raspbian, a Windows-based PC with an Ethernet port, an SD card reader for the PC (just required for the first time, either inbuilt or as an add-on SD card reader), and the RasPi (obviously!).

Assemble all of these on a neat table and just start your laptop without starting up the RasPi. I will run you through the step-by-step process. If you follow it, you'll have a working Internet connection provided from your PC to your RasPi with no added costs of Wi-Fi dongles.

Editing the command-line file of the RasPi

The first step is to edit the file that RasPi checks when it starts booting. Try inserting the SD card of the RasPi into your PC's SD card reader. Open **Explorer** (where all the drives are listed); there, you can find the removable media. You will be amazed to see that the partition is about 15 MB to 20 MB, but your card is actually 8 GB or 16 GB! Don't panic; it's just the boot space of RasPi. You will be able to see the multiple files on this media. We are interested in editing the cmdline.txt file. Just double-click on the file (or open it in standard Notepad), and you will be able to see the following startup commands:

```
dwc_otg.lpm_enable=0 console=ttyAMA0,115200 kgdboc=ttyAMA0,115200
console=tty1 root=/dev/mmcblk0p2 rootfstype=ext4 elevator=deadline
rootwait
```

You can change some settings by adding the static IP address of your RasPi at the end of the line (take a look at the following code). There is no need to understand the meaning of all of these parameters at this stage; I will introduce them when they will be useful.

```
dwc_otg.lpm_enable=0 console=ttyAMA0,115200 kgdboc=ttyAMA0,115200
console=tty1 root=/dev/mmcblk0p2 rootfstype=ext4 elevator=deadline
rootwait ip=169.254.0.2
```

In bold, you will see the static IP we have provided for the RasPi.

From now onwards, you'll always have to access your Pi using this IP address, when you access it from your PC.

 If you are a Linux or Mac OS X user, insert the SD card into the SD card reader. There will be two partitions visible. Open the boot partition and follow the same process explained to add the IP address to the cmdline.txt file.

Save the file, safely remove the SD card from the PC, and move on to the next step.

Turning on the RasPi

Now it's time to start and boot the RasPi by inserting the SD back into the RasPi. Establish an Ethernet connection between the RasPi and your PC before powering up the RasPi. You will see now multiple LEDs blinking on RasPi, stating that the Ethernet connection is being established and there is a transfer of data occurring between the PC and the RasPi. Check the working connection of PC and RasPi by entering ping 169.254.0.2 in Command Prompt (**Start Menu | Run | cmd.exe**). Note that we are using the same IP address as entered in the cmdline.txt file. It should give a response like this:

```
C:\Users\RUSHI>ping 169.254.0.2

Pinging 169.254.0.2 with 32 bytes of data:
Reply from 169.254.0.2: bytes=32 time=1ms TTL=64
Reply from 169.254.0.2: bytes=32 time<1ms TTL=64
Reply from 169.254.0.2: bytes=32 time<1ms TTL=64
Reply from 169.254.0.2: bytes=32 time<1ms TTL=64

Ping statistics for 169.254.0.2:
    Packets: Sent = 4, Received = 4, Lost = 0 (0% loss),
Approximate round trip times in milli-seconds:
    Minimum = 0ms, Maximum = 1ms, Average = 0ms
```

The ping command allows us to send the predefined size of packets to the host systems and expects them to be reflected back. The Lost = 0 section in the response shows that all the packets sent from the PC to the RasPi are reflected back and the connection is working.

For Linux and Mac OS X users, the connection can be verified by entering the `ping 169.254.0.2` command in the terminal. Enter the `ping` command in the same Command Prompt to get the Ethernet port IP address of your PC. Following this, enter `ipconfig` (`ifconfig` in the case of Mac OS X and Linux users) and note down the IP address of the LAN connection (Ethernet), which is `169.254.121.232` in the following screenshot:

```
Ethernet adapter Local Area Connection:

    Connection-specific DNS Suffix  . :
    Link-local IPv6 Address . . . . . : fe80::ed8e:2601:5396:79e8%10
    Autoconfiguration IPv4 Address. . : 169.254.121.232
    Subnet Mask . . . . . . . . . . . : 255.255.0.0
    Default Gateway . . . . . . . . . :
```

 You will be able to see these Ethernet IP addresses only if the RasPi is in the "powered on" state. Otherwise, you will see no IP address.

Changing cmdline.txt again to add the PC's Ethernet port IP address

Shut down the RasPi (`sudo poweroff`), remove the SD card, insert it back into your PC, and follow the *Editing the command-line file of RasPi* section. Add the noted IP address (in this case, it's `169.254.121.232`) at the end of the `cmdline.txt` file, as shown in the following code:

```
dwc_otg.lpm_enable=0 console=ttyAMA0,115200 kgdboc=ttyAMA0,115200
console=tty1 root=/dev/mmcblk0p2 rootfstype=ext4 elevator=deadline
rootwait ip=169.254.0.2 ::169.254.121.232
```

Here, the double colon (`::`) is the most important part to be put between the RasPi's IP address and your PC's IP address. Then, save the `cmdline.txt` file.

Sharing the Internet connection between your PC and an Ethernet connection

Turn on the RasPi after safely inserting the SD card back into RasPi and plugging in the micro USB adapter. On a Windows PC, you need to open **Network and Sharing Center** by navigating to **Control Panel | Change adapter settings**, right-clicking on the adapter where you are getting Internet connection, and going to its properties.

Look out for possible Internet connectivity on your PC through a wireless Internet connection.

Click on the **Sharing** tab. Keep the **Allow other network users to connect through this computer's Internet connection** option checked and click on **OK**. This setting changes the IP address of the Ethernet port of the PC; we need to reset it.

In the same window of the adapter settings, go to the properties of the **Local Area Network** connection (Ethernet), double-click on the **IPv4** settings, and click on **Obtain an IP Address Automatically** as well as **Obtain DNS server address automatically**.

 The IP address we provide for the RasPi may have a subnet class different from the network in your home. The interesting point is that this subnet class remains between the RasPi and PC. Don't panic if the Wi-Fi adapter of your PC is getting IPs in range of 192.x.x.x. This method still works, as Windows allows the Internet sharing between cross subnets. This is because we have enabled the Internet sharing and Ethernet settings as automatic. Therefore, it is clear that the Wi-Fi-to-PC (192.x.x.x) and PC-to-RasPi (169.254.x.x) scenarios work successfully.

Mac OS X user can follow the same steps by navigating to **Preferences** | **Sharing** and it would be very easy to follow the GUI.

Linux users can click on network menu in the top panel and navigate to **Edit Connections...** and then double click your wired connection and keep the wireless connection untouched. Navigate to the **IPv4 Settings** tab and select method: **Shared to other computers**.

Installing and opening the free SSH client on your PC

Secure Shell (SSH) is a cryptographic network protocol for secure data communication. It means remote command-line login and remote command execution between two networked computers. Here, we use it for the command-line login and remote command execution between the PC and the RasPi. A one-of-a-kind and free SSH client is PuTTY (www.putty.org) for Windows, and since it is an open source, you can download it for free. Run PuTTY on your Windows PC and change the settings as follows:

For Linux and Mac OS X users, there is no need to install the PuTTY client, as they can directly perform this task from their terminal window by the ssh pi@169.254.0.2 command.

1. In the **Host Name** textbox of PuTTY, provide the same IP address that you entered in cmdline.txt (which is 169.254.0.2, as per the example given in the previous section).

2. Following this, navigate to **Category** | **Connection** | **SSH** | **X11** and check the **Enable X11 forwarding** option.

3. In the left-side **Category** menu, click on **Session**, enter the session name in the **Saved Sessions** field, and save it so that you don't have to save the settings every time you connect the RasPi with the PC.

4. Double-click on the saved connection and enter the ID and password; you will get the CLI on the screen of your laptop. Now how do we check the working Internet connection on the RasPi?

5. The answer to the preceding question is simple; enter the following command to check the Internet connection:

```
ping -c 4 www.google.com
```

You should get the same response with 0 percent packet loss, and now you have a working Internet connection on your RasPi. All of this is one-time hard work; later on, if you just have to keep your settings unchanged, log in to PuTTY, and enjoy the Internet on the RasPi. For the first time, setting up is somewhat a long process, but you know you've saved almost 10 USD for a Wi-Fi dongle. Isn't that a great thing?

 Here is a beautiful tip: you can install Xming from http://www.straightrunning.com/XmingNotes/, which is an X Windows System Server. Once it is installed, run it and you should see that there is no window. Worry not because as soon as you magically input the command lxsession in PuTTY, you will see the entire desktop of the RasPi on your PC's screen. A program similar to Xming is VNC Viewer, which directly opens the desktop of the RasPi by entering the IP address of the RasPi without logging in from PuTTY. Amazing, right?

A crash course on Linux

Many authors and books will teach you the concepts of the Linux operating system, so I will just quickly introduce Linux here. You can refer to *Beginning Linux Programming 4th Edition*, *Wrox Publishing*, written by Neil Matthew and Richard Stones. For shell scripting, you can refer to *Linux Shell Scripting Cookbook Second Edition*, *Packt Publishing*, written by Shantanu Tushar and Sarath Lakshman. This operating system is mostly known for its non-user-friendliness to beginners! When users start using Linux, they often wonder, "Why is this operating system widely used and famous?" Linux is the biggest open source platform for hobbyists like us, and it allows us to modify the kernel of the operating system the way we want. Some advantages of using Linux include (and are certainly not restricted to) being free, stable, quick, and dependable under the **General Public License** (**GPL**).

You can build your own personalized operating system using Linux kernel distributions, for example, a Raspbian developed on the Debian flavor of Linux. We need our customized operating systems because if we develop our personalized hardware, we know how it should be programmed and we develop the drivers according to our needs on top of the Linux kernel. We will now go through the most powerful tool in any Linux operating system—the terminal.

The terminal and shell

The most important tool of the Linux operating system is the terminal—the CLI of Linux. Windows users may have already come across Command Prompt (as we used in the previous section) and Mac OS X users may be familiar with the terminal. Once you learn the commands of the terminal, the Linux world opens up to you. Using the terminal, you can easily interact with the operating system and its kernel, which indirectly connects you and enables you to access the hardware resources. Shell is a command language translator (or interpreter) that executes command input from user. Shell uses the Linux kernel to execute commands.

In the terminal, you can use many shells. One of the most common shells is called **Bourne-Again Shell** (**Bash**). Unless you get into the complicated programming of shell, which is also known as shell scripting, you may not feel the strengths and weaknesses of a particular shell. Most of the simple commands remain the same for different shells. Shell scripting is useful when you want to do postprocessing on your files in the same or different directories, modify the usual commands of Linux by your own way, or just print or execute a program. Shell scripts allow several commands that would otherwise be entered manually in a command-line interface to be executed automatically, without waiting for a user to trigger each stage of the sequence. This allows us to create the preconfigured file to execute the C programs for our sensors in the upcoming projects, which can really save our time while creating the projects.

In the RasPi, LXTerminal is the tool that is ultimately the terminal for RasPi. If you are using PuTTY instead of the desktop, PuTTY CLI is the terminal, ultimately!

Useful and frequently used Linux commands

Well, this can be a very long list if we introduce and explain all the commands of Linux. Even a separate book twice this size is not enough to completely illustrate all the functionalities of the commands. We will see the commands that are essential and will be used throughout this book. This list can be used as a reference, and it's necessary to understand these commands:

- `pi@raspberrypi~$`: It's now time to introduce the most commonly seen command. It welcomes you on the first login and every subsequent login to your RasPi. This shows your username and the hostname of the RasPi. Here, the username is `pi` and the hostname is `raspberrypi`.

- `sudo`: This is an abbreviation of Superuser DO. This command gives you all privileges of the superuser (root, the most powerful user) of Linux. It is used in concatenation with other commands such as `nano`, `su`, `chmod`, and so on. By writing the `sudo su` command, you can enter superuser mode, in which you can execute, delete, and create any kind of files in any folders. This really gives you superpowers in Linux!

 The `sudo` command can be dangerous if not used properly. It can be used to hack into the Linux systems, or this superpower can allow you to delete the entire kernel of Linux; keep in mind that next time, the PC won't boot!

- `man`: This is the command that shows the manual of the Linux commands and different other function definitions. Type `man sudo` and you will get all the details related to the `sudo` command.

- `pwd`: This is an abbreviation of the present working directory. This shows you the current directory you are working in. Type `pwd` and press the *Enter* key. This should display something like `/home/pi`.

- `ls`: This is a command used to list the files or search for some files contained in a particular directory. Just typing `ls` and pressing the *Enter* key will give you a list of all files contained in the system. The options with `ls` are `-a`, `-l`, and so on. Just type `man ls` followed by the *Enter* key to see the different options available with it.

- `cd`: This command stands for change directory. Just give a path followed by the `cd` command and you will be taken to that directory. For example, `cd /home/pi/python_games` moves you directly to the `python_games` folder, while `cd ..` takes one step out of a particular directory.

- `apt-get`: This is the package manager for any Debian-based Linux distribution. It allows you to install and manage new software packages on your RasPi. Once you have an Internet connection on the RasPi, type `sudo apt-get install <package-name>`. It will first download the package and then install the same package. To remove a package, just use `sudo apt-get remove <package-name>`. To update and upgrade the operating system, you can use `sudo apt-get update` and `sudo apt-get upgrade` respectively.

- `cp`: This is used to copy the file from one directory to another, for example, `cp /home/pi/python_games/gem1.png /home/pi/gem1.png` will copy the `gem1.png` file to the `/home/pi` folder from `/home/pi/python_games`. You can use the `mv` command instead of `cp` to move the file from one folder to another.

- `rm`: This removes the specified file (or directory when `rmdir` is used). For example, `rm gem1.png`.

 Here's an important warning: files deleted in this way are generally not restorable.

- `cat`: This lists the content of the file; for example, `cat example.sh` will display the content of the `example.sh` file.

- `mkdir`: This creates the new directory in the present working directory; for example, `mkdir packt` will create a directory named `packt` in the present working directory. Just use the `ls` command to check whether it has been created or not by checking the list.

- `startx`: This command provides RasPi users with the user interface for running a window session.

- `sudo shutdown -h`: This leads to terminate all the processes on the RasPi, whereas `sudo halt` stops the CPU from running mode and halts the OS. The `sudo poweroff` command safely turns off the RasPi module.

These are the most frequently used commands for the RasPi. If more are needed in your projects, you will be introduced to them where relevant.

Let's create our first shell file:

1. Type `sudo nano example.sh` in the CLI of your RasPi (you can use PuTTY or the terminal on your PC and connect the RasPi through Ethernet connection with your PC). Just type the following code in the nano text editor:

   ```
   echo hello world
   echo this is my first shell program
   ```

2. Press *Ctrl* + *X* to exit and press *Y* to confirm the exit while also saving the file. The `echo` command simply displays the text on the screen of the terminal when executed; this is similar to the `printf` command in C, but is really simple compared to it, right?

3. Now enter the `sudo chmod +x example.sh` command in the terminal to provide execution permissions on the `example.sh` file.

4. Execute the shell program by just typing `./example.sh` (`./` means a dot followed by a forward slash, which makes the shell execute the filename that is after the forward slash).

Notice that this is very short introduction to shell, and now you will learn the useful commands that will be used throughout the book to create the projects.

Installing useful libraries

I compare this section to an ice cream with chocolate sauce (yummy! ssrupp!). If you have a vanilla ice cream in your hand, you can enjoy the ice cream, but once you pour chocolate sauce on that, it becomes more delicious, doesn't it? Adding and installing libraries in the RasPi is the same scenario. The RasPi is amazing with the added libraries, which can give you the functionalities you want, whether it is on the GPIO or on the camera port. A library is a particular set of functions that gives you easiness while writing the programs.

Step by step, we will install the useful libraries. To install the libraries, all you need is an Internet connection on the RasPi via PuTTY, as explained in previous sections.

Before installing any libraries, verify that your operating system has the latest update. Always check for upgrades and updates by entering these commands:

```
sudo apt-get update
sudo apt-get upgrade
```

Here, we update the RasPi to provide information on the latest package versions and dependencies. All the repositories will get information about their latest packages and to resynchronize. In the next step, upgrade will fetch new versions of packages according to the list provided in the update list. This process will take time, depending on the size of the update and the quality of the Internet connection.

git-core

git is a code management system used for collaborative work among programmers across the world, and it makes tracing change in the code easy. You will find many libraries and projects on git. If you know the source repository, you can directly get the library using git-core. Install git-core using this command:

```
sudo apt-get install git-core
```

wiringPi

The wiringPi library is created by Gordon, written in C, and provides you with support to extend your C programs to control the GPIOs. You can easily download (which will need the Internet connection shared on the RasPi) this library from Gordon's git core profile by typing the following command:

```
git clone git://git.drogon.net/wiringPi
```

The RasPi then downloads the library and creates a folder in the root directory. Use the cd wiringPi command to change the directory and go to the wiringPi directory. The next command to be entered is git pull origin, which fetches the latest version, and then we are ready to build the script using the ./build command.

Now, once the build process is done, we are ready to use the wiringPi library in any C program we write in the future. To check whether this particular library is working perfectly, enter these commands: gpio -v and gpio readall. This will convince you that you have installed it correctly. In *Chapter 2*, *Meeting the World of Electronics*, you will learn how to use wiringPi in shell script and the C language.

python-gpio

The latest distribution of the RasPi comes with `python-gpio` installed, but this library will be necessary for those who have an old distribution installed.
The `python-gpio` library allows you to easily access and control the GPIO pins while running the Python script. This library can be downloaded from the Python organization's website, but we will install it using the LXTerminal or PuTTY. Let's proceed by downloading the `.tar` file:

```
wget https://pypi.python.org/packages/source/R/RPi.GPIO/RPi.GPIO-
0.5.7.tar.gz
```

If you aren't aware, let me tell you that a TAR file is a kind of bundled file used to make the download compact and easy. We need to extract the downloaded file in a directory:

```
tar -xf RPi.GPIO-0.5.7.tar.gz
```

Let's rename this folder for ease of use; use this command:

```
mv RPi.GPIO-0.5.7 python_gpio
```

Move to the `python-gpio` directory to install the library using this command:

```
cd python_gpio
```

> While writing a command, you can use the *Tab* key, which provides an autocompletion feature. For example, while writing the `cd pyth` command, press *Tab*. This will autocomplete the command, which will save the time spent on long filenames. Pressing the *Tab* key twice will give you a list of the available commands or filenames.

Now, we will install this library:

```
sudo python setup.py install
```

There is a possibility that it gives you a response that the library is already installed with the latest version. After this process, if you want to remove the downloaded file, you can use the `rm` command and remove it. Finally, one more library that provides support for the Python **Serial Peripheral Interface (SPI)** protocol on GPIO is `spidev`. You can install it using the following command. You can refer to the *Serial Peripheral Interface* section of *Chapter 2, Meeting the World of Electronics*, to learn more on SPI protocol. We will be using the SPI protocol in *Chapter 6, Uploading Data Online – Spreadsheets, Mobile, and E-mails*, when we build the sensor station project to send sensor data on web pages.

```
sudo pip install spidev
```

> There are many libraries available, but we will install them later in the upcoming chapters, when the need arises. It is simple to install the libraries, why wouldn't it be? Linux rocks!

Be ready with Python and C

We'll use Python because it is a very simple, yet powerful, language and is easy to write and read because of its indentation and standard English keywords.

> There are two major versions available and there is a current debate on Python 2 versus Python 3. You can read it at https://wiki.python.org/moin/Python2orPython3.
>
> This book will mostly use Python 2.7.x. If you are beginner and want to learn Python, I advise you to go with Python 3; there is not much difference between the two, but there are noticeable differences and you will observe them.

The C programming language offers ample benefits when developing the projects using already available libraries, such as wiringPi, which can give you full control of GPIO pins. If you have previously developed a project on C, you can integrate the wiringPi functions and get the same functionalities as your previous project. Also, you can simultaneously use GPIO.

Let's play around with both the languages; this will not give you the whole idea of the programming, but it will give you a good start and will create interest. We will see both the languages one by one.

Writing and executing the Python program

When you use the RasPi, the Python is already an installed component. In the Linux CLI, you can just type python and the Python CLI will wait for you to enter the commands. Just type print "this is my first program in Python" and press *Enter*. Voilà! You have executed a command of Python. This will not allow you to write full-length code directly now, so what to do if you want to write a long code? There is a better way than this, and we will use that throughout the book.

Type `sudo nano example1.py` and you will observe the nano text editor on the screen. Then type the following code:

```
name = "World"
name = "Hello " + name
print name
for i in range(3):
    print "Whoa"
import this
```

Now press *Ctrl + X* and then press *Y* to save the changes. You will be back to the Linux CLI. Now type `python example1.py`. The Python program will be compiled and the output will be displayed in the same window. One thing you should notice is that indentation is very necessary in Python. Remove the indentation before the `print "whoa"` script and then execute the program; you will find an error of indentation. In the loops, special care for inserting indentation should be taken while writing the code. This makes the programs easy to read for people other than programmers.

Writing and executing the C program

You should know that the most powerful language existing today is C, and it allows us to fulfil all our coding needs. The C language is very common and is an essential language. Let's go through the procedure of executing a C program, which is almost similar to executing Python programs.

Type `sudo nano example2.c` in the LXTerminal or PuTTY. Then you can type any C code you know, or on a beginner basis, you can try this code:

```
#include<stdio.h>
int i;
int main()
{
for(i=0;i<3;i++)
printf("Harder you work, Luckier you get");
return 0;
}
```

Press *Ctrl + X* and press *Y* to save the changes. Now it's our turn to compile and execute the program. The compiler for C programs is always included in the Linux distribution, which can be `cc` or `gcc`. Type this command to compile the C program:

```
gcc -o example2 example2.c
```

In `-o example2`, the `example2` part will be the name of the output file and the `example2.c` part is the file we saved after writing the C program. Press *Enter* and check the errors. Correct it by typing `sudo nano example2.c` and solve it (if any error occurs). Once it is successfully compiled, type the `ls` command to check whether the output file has been created. The output filename will be `example2`. You can now type `./example2` to execute the compiled code.

These processes are really helpful in creating sensor projects, and once you practice more codes, it will be easy for you to understand the process.

Practice makes you perfect

This section includes some practice problems, which should be exercised with shell, C, and Python. The reason behind this practice is that it will make you stronger in understanding the problems and logic of programs, which can really help you to easily make the sensor project. This practice will not cover or give you the idea of entire language or script, but will make you comfortable enough to understand the codes used in the next chapters.

I advise you to connect the Raspberry Pi through an Ethernet cable with your PC, and use the methods stated in the preceding sections to execute the programs. You can take help of the Internet (`www.cprogramming.com/quiz/`) to understand the logic and the syntax. The problem statements shown here should be attempted in all three languages/scripts, which will give you enough idea to work with scripts and languages:

- Write a program to get all Armstrong numbers below 1000. Note that among three digit numbers, an Armstrong number is equal to the sum of the cubes of its digits. For example, 153 is an Armstrong number because $153 = 1^3 + 5^3 + 3^3$.

- Convert the temperature value from degrees Celsius to degrees Fahrenheit and vice versa. Ask user to get the value and decide whether they are entering it in Celsius or Fahrenheit. Show a warning message if the temperature is above 38 degrees Celsius or 100 degrees Fahrenheit.

- Create a calculator that has all the basic functionalities, such as addition, subtraction, division, and multiplication. Ask the user to select the function they want. Show an error if they divide anything by zero.

- Get a time value from the clock, attach to a `Time is` string, and display the current time, for example, `Time is 17-Oct-14 10:18:22AM`.

The skills acquired by performing these exercises will allow you to better understand the projects in the upcoming chapters. You can expect an easy programming level in upcoming chapters. These chapters will focus more on Python and C programs. Hence, more practice on programs will help you gain a better understanding of the language and increase logical thinking.

Summary

In this chapter, you learned about the different connectors and functionalities available on the Raspberry Pi board. We successfully installed the operating system on the RasPi and shared an Internet connection with it. After these processes, you learned the basic Linux commands and a glance of the Linux terminal and shell scripting, which will be used frequently while developing applications and projects. Then we installed the useful libraries (in the same way as we add a chocolate topping on top of a vanilla ice cream). A brief introduction to compiling and executing C and Python programs was given to kick-start work on the Raspberry Pi.

I am sure that you will solve the problems stated at the end of this chapter to get an idea of how code works. This will help a lot in the upcoming chapters.

In the next chapter, you will be learning the basics of electronics so that you can easily develop the projects. These basics are essential for interfacing the sensors. You will also learn how sensors communicate with the Raspberry Pi. We will run simple codes to drive LEDs on GPIO pins.

2
Meeting the World of Electronics

You can't spend even a day without electronics, can you? Electronics is everywhere, from your toothbrush to cars and in aircrafts and spaceships too. In this chapter, we will go through the fundamental concepts of electronics that will be useful while building our projects so that one day we can make our own products and amaze the world. This chapter will help you understand the concepts of electronics that can be very useful while working with the RasPi.

You might have read many electronics-related books, and they might have bored you with concepts when you really wanted to create or build projects. I believe that there must be a reason for explanations being given about electronics and its applications. Hence this chapter provides basic explanations of various terminologies in electronics and their usefulness in the projects.

Once you know about the electronics, we will walk through the communication protocols and their uses with respect to communication among electronic components and different techniques to do it. Useful tips and precautions are listed before starting to work with GPIOs on the RasPi. Then, you will understand the functionalities of GPIO and blink the LED using shell, Python, and C code.

In this chapter, you will learn the following topics:

- Basic electronics terminologies and some fundamentals
- How electronic components communicate with each other using the UART, SPI, and I2C protocols
- GPIO essentials
- GPIO port functionality and glowing LED using shell scripting, Python, and C language

First let's cover some of the fundamentals of electronics.

Basic terminologies of electronics

There are numerous terminologies used in the world of electronics. From the hardware to the software, there are millions of concepts that are used to create astonishing products and projects. You already know that the RasPi is a single-board computer that contains plentiful electronic components built in, which makes us very comfortable to control and interface the different electronic devices connected through its GPIO port. In general, when we talk about electronics, it is just the hardware or a circuit made up of several **Integrated Circuits** (ICs) with different resistors, capacitors, inductors, and many more components. But that is not always the case; when we build our hardware with programmable ICs, we also need to take care of internal programming (the software). For example, in a microcontroller or microprocessor, or even in the RasPi's case, we can feed the program (technically, permanently burn/dump the programs) into the ICs so that when the IC is powered up, it follows the steps written in the program and behaves the way we want. This is how robots, your washing machines, and other home appliances work. All of these appliances have different design complexities, which depends on their application. There are some functions, which can be performed by both software and hardware. The designer has to analyze the trade-off by experimenting on both; for example, the decoder function can be written in the software and can also be implemented on the hardware by connecting logical ICs. The developer has to analyze the speed, size (in both the hardware and the software), complexity, and many more parameters to design these kinds of functions. The point of discussing these theories is to get an idea on how complex electronics can be. We need not look deeper into electronics now, but let's start with the important entities listed in the following sections. It is very important for you to know these terminologies because you will need them frequently while building the RasPi projects.

Voltage

Who discovered voltage? Okay, that's not important now, let's understand it first. The basic concept follows the physics behind the flow of water. Water can flow in two ways; one is a waterfall (for example, from a mountain top to the ground) and the second is forceful flow using a water pump. The concept behind understanding voltage is similar. Voltage is the potential difference between two points, which means that a voltage difference allows the flow of charges (electrons) from the higher potential to the lower potential. To understand the preceding example, consider lightning, which can be compared to a waterfall, and batteries, which can be compared to a water pump. When batteries are connected to a circuit, chemical reactions within them pump the flow of charges from the positive terminal to the negative terminal. Voltage is always mentioned in volts (V). The AA battery cell usually supplies 3V. By the way, the term voltage was named after the great scientist Alessandro Volta, who invented the voltaic cell, which was then known as a battery cell.

Current

Current is the flow of charges (electrons). Whenever a voltage difference is created, it causes current to flow in a fixed direction from the positive (higher) terminal to the negative (lower) terminal (known as conventional current). Current is measured in amperes (A). The electron current flows from the negative terminal of the battery to the positive terminal. To prevent confusion, we will follow the conventional current, which is from the positive terminal to the negative terminal of the battery or the source.

Resistor

The meaning of the word "resist" in the Oxford dictionary is "to try to stop or to prevent." As the definition says, a resistor simply prevents the flow of current. When current flows through a resistor, there is a voltage drop in it. This drop directly depends on the amount of current flowing through resistor and value of the resistance. There is a formula used to calculate the amount of voltage drop across the resistor (or in the circuit), which is also called as the Ohm's law ($V = I * R$). Resistance is measured in ohms (Ω). Let's see how resistance is calculated with this example: if the resistance is 10Ω and the current flowing from the resistor is 1A, then the voltage drop across the resistor is 10V. Here is another example: when we connect LEDs on a 5V supply, we connect a 330Ω resistor in series with the LEDs to prevent blow-off of the LEDs due to excessive current. The resistor drops some voltage in it and safeguards the LEDs. We will extensively use resistors to develop our projects.

Capacitor

A resistor dissipates energy in the form of heat. In contrast to that, a capacitor stores energy between its two conductive plates. Often, capacitors are used to filter voltage supplied in filter circuits and to generate clear voice in amplifier circuits. Explaining the concept of capacitance will be too hefty for this book, so let me come to the main point: when we have batteries to store energy, why do we need to use capacitors in our circuits? There are several benefits of using a capacitor in a circuit. Many books will tell you that it acts as a filter or a surge suppressor, and they will use terms such as power smoothing, decoupling, DC blocking, and so on. In our applications, when we use capacitors with sensors, they hold the voltage level for some time so that the microprocessor has enough time to read that voltage value. The sensor's data varies a lot. It needs to be stable as long as a microprocessor is reading that value to avoid erroneous calculations. The holding time of a capacitor depends on an RC time constant, which will be explained when we will actually use it.

Open circuit and short circuit

Now, there is an interesting point to note: when there is voltage available on the terminal but no components are connected across the terminals, there is no current flow, which is often called an open circuit. In contrast, when two terminals are connected, with or without a component, and charge is allowed to flow, it's called a short circuit, connected circuit, or closed circuit.

Here's a warning for you: do not short (directly connect) the two terminals of a power supply such as batteries, adaptors, and chargers. This may cause serious damages, which include fire damage and component failure. If we connect a conducting wire with no resistance, let's see what Ohm's law results in: $R = 0\Omega$ then $I = V/0$, so $I = \infty A$. In theory, this is called infinite (uncountable), and practically, it means a fire or a blast!

Series and parallel connections

In electrical theory, when the current flowing through a component does not divide into paths, it's a series connection. Also, if the current flowing through each component is the same then those components are said to be in series. If the voltage across all the components is the same, then the connection is said to be in parallel. In a circuit, there can be combination of series and parallel connections. Therefore, a circuit may not be purely a series or a parallel circuit. Let's study the circuits shown in the following diagram:

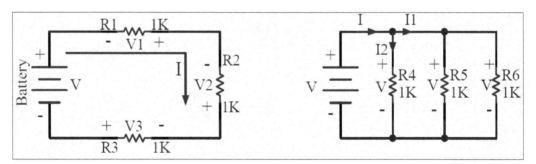

Series and parallel connections

At the first glance, this figure looks complex with many notations, but let's look at each component separately. The figure on the left is a series connection of components. The battery supplies **voltage (V)** and **current (I)**. The direction of the current flow is shown as clockwise. As explained, in a series connection, the current flowing through every component is the same, but the voltage values across all the components are different. Hence, $V = V1 + V2 + V3$. For example, if the battery supplies 12V, then the voltage across each resistor is 4V. The current flowing through each resistor is 4 mA (because $V = IR$ and $R = R1 + R2 + R3 = 3K$).

The figure on the right represents a parallel connection. Here, each of the components gets the same voltage but the current is divided into different paths. The current flowing from the positive terminal of the battery is I, which is divided into I1 and I2. When I1 flows to the next node, it is again divided into two parts and flown through R5 and R6. Therefore, in a parallel circuit, $I = I1 + I2$. The voltage remains the same across all the resistors. For example, if the battery supplies 12V, the voltage across all the resistors is 12V but the current through all the resistors will be different. In the parallel connection example, the current flown through each circuit can be calculated by applying the equations of current division. Give it a try to calculate!

When there is a combination of series and parallel circuits, it needs more calculations and analysis. Kirchhoff's laws, nodes, and mesh equations can be used to solve such kinds of circuits. All of that is too complex to explain in this book; you can refer any standard circuits-theory-related books and gain expertise in it.

Kirchhoff's current law: At any node (junction) in an electrical circuit, the sum of currents flowing into that node is equal to the sum of currents flowing out of that node.

Kirchhoff's voltage law: The directed sum of the electrical potential differences (voltage) around any closed network is zero.

Pull-up and pull-down resistors

Pull-up and pull-down resistors are one of the important terminologies in electronic systems design. As the title of this section says, there are two types of pulling resistors: pull-up and pull-down. Both have the same functionality, but the difference is that pull-up resistor pulls the terminal to the voltage supplied and the pull-down resistor pulls the terminal to the ground or the common line. The significance of connecting a pulling resistor to a node or terminal is to bring back the logic level to the default value when no input is present on that particular terminal. The benefit of including a pull-up or pull-down resistor is that it makes the circuitry susceptible to noise, and the logic level (1 or 0) cannot be changed from a small variation in terms of voltages (due to noise) on the terminal. Let's take a look at the example shown in the following figure. It shows a pull-up example with a NOT gate (a NOT gate gives inverted output in its OUT terminal; therefore, if logic one is the input, the output is logic zero). We will consider the effects with and without the pull-up resistor. The same is true for the pull-down resistor.

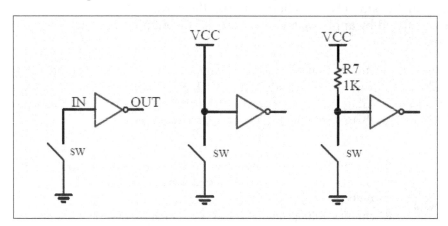

Connection with and without pull-up resistors

In general, logic gates have high impedance at their input terminal, so when there is no connection on the input terminal, it is termed as floating. Now, in the preceding figure, the leftmost connection is not recommended because when the switch is open (OFF state), it leaves the input terminal floating and any noise can change the input state of the NOT gate. The reason of the noise can be any. Even the open terminals can act as an antenna and can create noise on the pin of the NOT gate. The circuit shown in the middle is a pull-up circuit without a resistor and it is highly recommended not to use it. This kind of connection can be called a pull-up but should never be used. When the switch is closed (ON state), the VCC gets a direct path to the ground, which is the same as a short circuit. A large amount of current will flow from VCC to ground, and this can damage your circuit.

The rightmost figure shows the best way to pull up because there is a resistor in which some voltage drop will occur. When the switch is open, the terminal of the NOT gate will be floated to the VCC (pulled up), which is the default. When the switch is closed, the input terminal of the NOT gate will be connected to the ground and it will experience the logic zero state. The current flowing through the resistor will be nominal this time. For example, if *VCC = 5V, R7 = 1K, and I = V/R*, then I = 5mA, which is in the safe region. For the pull-down circuit example, there can be an interchange between the switch and a resistor. The resistor will be connected between the ground and the input terminal of the NOT gate. When using sensors and ICs, keep in mind that if there is a notation of using pull-ups or pull-downs in datasheets or technical manuals, it is recommended to use them wherever needed. In the next section, we will use the pull-up resistor in one of the communication protocols.

Communication protocols

It has been a lot theory so far. The previous section was meant to give you an understanding of some useful concepts of electronics. There can be numerous components, including ICs and digital sensors, as peripherals of a microprocessor. There can be a large amount of data with the peripheral devices, and there might be a need to send it to the processor. How do they communicate? How does the processor understand that the data is coming into it and that it is being sent by the sensor? There is a serial, or parallel, data-line connection between ICs and a microprocessor. Parallel connections are faster than the serial one but are less preferred because they require more lines, for example, 8, 16, or more than that. A PCI bus can be an example of a parallel communication. Usually in a complex or high-density circuit, the processor is connected to many peripherals, and in that case, we cannot have that many free pins/lines to connect an additional single IC. Serial communication requires up to four lines, depending on the protocol used. Still, it cannot be said that serial communication is better than parallel, but serial is preferred when low pin counts come into the picture. In serial communication, data is sent over frames or packets. Large data is broken into chunks and sent over the lines by a frame or a packet. Now, what is a protocol? A protocol is a set of rules that need to be followed while interfacing the ICs to the microprocessor, and it's not limited to the connection. The protocol also defines the data frame structures, frame lengths, voltage levels, data types, data rates, and so on. There are many standard serial protocols such as UART, FireWire, Ethernet, SPI, I2C, and more. The RasPi 1 models B, A+, B+, and the RasPi 2 model B have one SPI pin, one I2C pin, and one UART pin available on the expansion port. We will see these protocols one by one.

UART

UART is a very common interface, or protocol, that is found in almost every PC or microprocessor. UART is the abbreviated form of Universal Asynchronous Receiver and Transmitter. This is also known as the RS-232 standard. This protocol is full-duplex and a complete standard, including electrical, mechanical, and physical characteristics for a particular instance of communication. When data is sent over a bus, the data levels need to be changed to suit the RS-232 bus levels. Varying voltages are sent by a transmitter on a bus. A voltage value greater than 3V is logic zero, while a voltage value less than -3V is logic one. Values between -3V to 3V are called as undefined states. The microprocessor sends the data to the **transistor-transistor logic** (TTL) level; when we send them to the bus, the voltage levels should be increased to the RS-232 standard. This means that to convert voltage from logic levels of a microprocessor (0V and 5V) to these levels and back, we need a level shifter IC such as MAX232. The data is sent through a DB9 connector and an RS-232 cable. Level shifting is useful when we communicate over a long distance.

What happens when we need to connect without these additional level shifter ICs? This connection is called a NULL connection, as shown in the following figure. It can be observed that the transmit and receive pins of a transmitter are cross-connected, and the ground pins are shared. This can be useful in short-distance communication. In UART, it is very important that the baud rates (symbols transferred per second) should match between the transmitter and the receiver. Most of the time, we will be using 9600 or 115200 as the baud rates. The typical frame of UART communication consists of a start bit (usually 0, which tells receiver that the data stream is about to start), data (generally 8 bit), and a stop bit (usually 1, which tells receiver that the transmission is over).

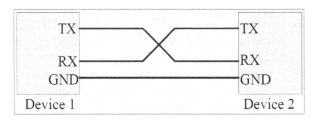

Null UART connection

The following figure represents the UART pins on the GPIO header of the
RasPi board. Pin 8 and 10 on the RasPi GPIO pin header are transmit and
receive pins respectively.

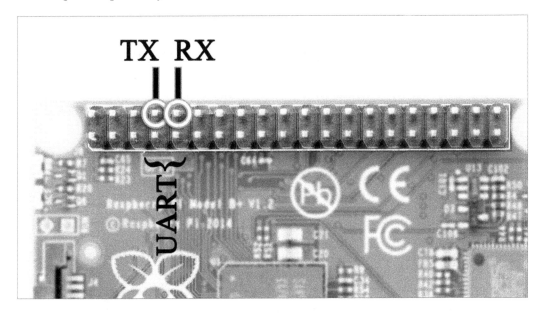

Many sensors do have the UART communication protocol enabled on their
output pins. Sensors such as gas sensors (MQ-2) use UART communication
to communicate with the RasPi. Another sensor that works on UART is the
nine-axis motion sensor from LP Research (LPMS-UARTL), which allows you to
make quadcopters on your own by providing a three-axis gyroscope, three-axis
magnetometer, and three-axis accelerometer. The TMP104 sensor from Texas
instruments comes with UART interface digital temperature sensors. Here, the
UART allows daisy-chain topology (in which you connect one's transmit to the
receive of the second, the second's transmit to the third's receive, and so on up to
eight sensors). In a RasPi, there should be a written application program with the
UART driver in the Python or C language to obtain the data coming from a sensor.

Serial Peripheral Interface

The **Serial Peripheral Interface (SPI)** is a full-duplex, short-distance, and single-master protocol. Unlike UART, it is a synchronous communication protocol. One of the simple connections can be the single master-slave connection, which is shown in the next figure. There are usually four wires in total, which are clock, **Master In Slave Out (MISO)**, **Master Out Slave In (MOSI)**, and **chip select (CS)**. Have a look at the following image:

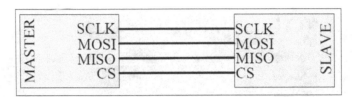

Simple master-slave SPI connections

The master always initiates the data frame and clock. The clock frequencies can be varied from the master according to the slave's performance and capabilities. The clock frequency varies from 1 MHz to 40 MHz, and higher too. Some slave devices trigger on active low input, which means that whenever the logic zero signal is given by the master to slave on the CS pin, the slave chip is turned ON. Then it accepts the clock and data from master. There can be multiple slaves connected to a master device. To connect multiple slaves, we need additional CS lines from the master to be connected with the slaves. This can be one of the disadvantages of the SPI communication protocol, when slaves are increased. There is no slave acknowledgement sent to the master, so the master sends data without knowing whether the slave has received it or not. If both the master and the slave are programmable, then during runtime (while executing the program), the master and slave actions can be interchanged. For the RasPi, we can easily write the SPI communication code in either Python or C. *Chapter 5, Using an ADC to Interface any Analog Sensor with the Raspberry Pi*, comes with the use of the SPI protocol, in which we are going to interface a sensor station made by ourselves to log data. This logged data will be uploaded to the Internet and also sent to your own e-mail IDs. Interesting isn't it? The location of the SPI pins on RasPi 1 models A+ and B+ and RasPi 2 model B can be seen in the following diagram. This diagram is still valid for RasPi 1 model B:

Inter-Integrated Circuit

Inter-Integrated Circuit (I2C) is a protocol that works with two wires and it is a half-duplex (a type of communication where whenever the sender sends the command, the receiver just listens and cannot transmit anything; and vice versa), multimaster protocol that requires only two wires, known as data (SDA) and clock (SCL). The I2C protocol is patented by Philips, and whenever an IC manufacturer wants to include I2C in their chip, they need a license. Many of the ICs and peripherals around us are integrated with the I2C communication protocol. The lines of I2C (SDA and SCL) are always pulled up via resistors to the input voltage. The I2C bus works in three speeds: high speed (3.4 MBps), fast (400 KBps), and slow (less than 100 KBps). It is heard that the I2C communication is done up to 45 feet, but it's better to keep it under 10 feet.

Each I2C device has an address of 7 to 10 bits; using this address, the master can always connect and send data meant for that particular slave. The slave device manufacturer provides you with the address to use when you are interfacing the device with the master. Data is received at every slave, but only that slave can take the data for which it is made. Using the address, the master reads the data available in the predefined data registers in the sensors, and processes it on its own.

The general setup of an I2C bus configuration can be done as shown in the following diagram:

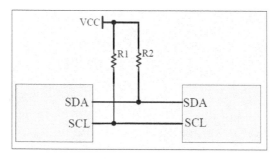

I2C bus interface

There are 16 x 2 character LCD modules available with the I2C interface in stores; you can just use them and program the RasPi accordingly. Usually, the LCD requires 8/4 wire parallel data bits, reset, read/write, and enable pins. The I2C pins are represented in the following image, and they can be located in the same place on all the RasPi models:

The I2C protocol is the most widely used protocol among all when we talk about sensor interfacing. Silicon Labs' Si1141 is a proximity and brightness sensor that is nowadays used in mobile phones to provide the auto-brightness and near proximity features. You can purchase it and easily interface it with the RasPi. SHT20 from Sensirion also comes with the I2C protocol, and it can be used to measure temperature and humidity data. Stepper motor control can be done using I2C-based controllers, which can be interfaced with the RasPi. The most amazing thing is that if you have all of these sensors, then you can tie them to a single I2C, but with RasPi you can get the data! You will get to know in the next section that we cannot use that many GPIOs on RasPi 1 model B or B+ and RasPi 2 model B. Therefore, the modules with the I2C interface are available for low-pin-count devices. This is why serial communication is useful.

These protocols are mostly used with the RasPi. The information given here about them is not that detailed, as numerous pages can be written on these protocols, but while programming the RasPi, this much information can help you build the projects.

Useful tips and precautions

Before we discuss the practicality side of the RasPi, let's look at the precautions and tips when working with RasPi. You need to read this, as you are now going to work with the GPIOs of the RasPi. This section will give you generalized tips and warnings to keep in mind when working with the RasPi:

- Avoid touching the electronic components on the RasPi, as even a small amount of sweat or a static charge from your body can spoil the board. The components on the board are so small that it can be affected by a very small amount of sweat in your hand, and by water too. Touch it from the corners, or always use a casing to cover it. There are a plenty of these available in e-stores.

- Take care when putting the RasPi on the table. If there is a small metal part (cut pieces of wires) or the table itself is made of a metal, it can short the connections on the RasPi.

- Never connect any device that provides voltage higher than 3.3V to the RasPi's GPIO pins. In *Chapter 3*, *Measuring Distance Using Ultrasonic Sensors*, we will face this scenario and you will learn how it can be overcome. Whenever you think of connecting any device to the RasPi, get all the details of the power ratings of that device. Please avoid connecting the 5V supply line of the RasPi by connecting the jumper wire to the GPIOs.

- When using low-power LEDs, it is good practice to include the resistor (270 to 330Ω will be good) in series with the positive wire of the LED. Please do not randomly plug the power sources and wires into your RasPi. It is advised that you should not short the two pins on the board directly.

- Do not try to connect components that require much power to your RasPi; LEDs are okay, DC motors are not. If you want to interface the motors, you need an additional motor driver add-on circuit with RasPi. Take a look at the Pibrella board for it.

Let's look into something more practical now.

Understanding the GPIO port

You will find working with GPIO very interesting! You already know from the first chapter that the GPIO pins are the configurable pins of a processor and if you will closely see the RasPi board, the GPIO functionality is brought out on board pin-out header from the processor in such a way that the GPIO status can be changed and also be read during the runtime. That is what we are going to do in this section. While programming, you will notice that the RasPi's GPIO has two modes: board mode and BCM mode. Board mode can be seen as the pin numbers physically seen on the board, which are internally connected to the processor. As the processor has numerous pins and GPIOs available, the processor pin number and the board header pin number will always be different. For example, the processor has internally assigned the GPIO 17 designation on its own pin, while on the RasPi board, a connector will have number 11 of the pin. Let's see the available GPIO pins and their functionality on RasPi 1 models B, A+, and B+ and RasPi 2 model B in the following tables. The first table shows the GPIO out designations for RasPi 1 model B. For RasPi 2 model B and RasPi 1 models A+ and B+, there are 40 pins on the header. Therefore, it can be said that the RasPi has two types of connectors, one with 26 pins and the latest models with 40-pin headers. All of these RasPi models have the same number of SPI, UART, and I2C interfaces available as pin-outs on the board.

Physical/RasPi names (Board)				Broadcom names (BCM)			
Left		Right		Left		Right	
Pin Function	Pin No.	Pin No.	Pin Function	Pin Function	Pin No.	Pin No.	Pin Function
3.3V	1	2	5V	3.3V	1	2	5V
I2C0 SDA	3	4	DNC	I2C0 SDA	3	4	DNC
I2C0 SCL	5	6	GND	I2C0 SCL	5	6	GND
GPIO 7	7	8	UART TX	GPIO 4	7	8	UART TX
DNC	9	10	UART RX	DNC	9	10	UART RX
GPIO 11	11	12	GPIO 12	GPIO 17	11	12	GPIO 18
GPIO 13	13	14	DNC	GPIO 21	13	14	DNC
GPIO 15	15	16	GPIO 16	GPIO 22	15	16	GPIO 23
DNC	17	18	GPIO 18	DNC	17	18	GPIO 24
SPI MOSI	19	20	DNC	SPI MOSI	19	20	DNC
SPI MISO	21	22	GPIO 22	SPI MISO	21	22	GPIO 25
SPI SCLK	23	24	SPI CE0	SPI SCLK	23	24	SPI CE0
DNC	25	26	SPI CE1	DNC	25	26	SPI CE1

If we compare this pin-out with RasPi 1 model B+, we can see that every pin up to number 26 is the same. Every pin on this port acts as an access point to communicate with the RasPi. Let's see the pins one by one. Pins 1 and 2 are the power supply, if you need low-power sensors or some peripheral needs to be powered by these pins.

Be careful when using the 3.3V and 5V pins. Never connect both the voltage supply pins with each other directly or the 5V line to any of the GPIOs directly. You may think that you are giving logic high for project purposes, but the RasPi is made to take only 3.3V and not 5V, so choose the peripherals that can work on 3.3V levels. With the peripherals working on 5V, you need some protection circuit in between (such as diode circuitry), and then you can connect to the GPIO line.

Now, pins 3 and 5 are the pins for the I2C communication protocol. You might think that you will need pull-up resistors as discussed in the previous sections with I2C, but luckily, these pins are already pulled up with the resistors (1.8 kΩ) on board with 3.3V. You need not worry about the pulling up for interfacing of the sensors.

When not using these pins for I2C, we can use them for a GPIO, and internal pull-up resistors can help us interface switches very gracefully. So, the switch directly shorts this pin to the ground, and you can read the switch value in a piece of code written in the RasPi. Voilà! The following table shows the features of the RasPi model B and the RasPi 1 A+ and B+ GPIO pins:

Physical / RasPi name(Board)					Broadcom names (BCM)			
Left		Right			Left		Right	
Pin Function	Pin No.	Pin No.	Pin Function		Pin Function	Pin No.	Pin No.	Pin Function
3.3V	1	2	5V		3.3V	1	2	5V
I2C0 SDA	3	4	5V		I2C0 SDA	3	4	5V
I2C0 SCL	5	6	GND		I2C0 SCL	5	6	GND
GPIO 7	7	8	UART TXD		GPIO 4	7	8	UART TXD
GND	9	10	UART RXD		GND	9	10	UART RXD
GPIO 11	11	12	GPIO 12		GPIO 17	11	12	GPIO 18
GPIO 13	13	14	GND		GPIO 21	13	14	GND
GPIO 15	15	16	GPIO 16		GPIO 22	15	16	GPIO 23
3.3V	17	18	GPIO 18		3.3V	17	18	GPIO 24
SPI MOSI	19	20	GND		SPI MOSI	19	20	GND
SPI MISO	21	22	GPIO 22		SPI MISO	21	22	GPIO 25
SPI SCLK	23	24	SPI CE0		SPI SCLK	23	24	SPI CE0
GND	25	26	SPI CE1		GND	25	26	SPI CE1
ID_SD	27	28	ID_SC		ID_SD	27	28	ID_SC
GPIO 29	29	30	GND		GPIO 5	29	30	GND
GPIO 31	31	32	GPIO 32		GPIO 6	31	32	GPIO 12
GPIO 33	33	34	GND		GPIO 13	33	34	GND
GPIO 35	35	36	GPIO 36		GPIO 19	35	36	GPIO 16
GPIO 37	37	38	GPIO 38		GPIO 26	37	38	GPIO 20
GND	39	40	GPIO 40		GND	39	40	GPIO 21

Pins such as 7, 11, 13, 15, and so on act as GPIO pins. RasPi 1 model B has eight dedicated GPIO pins available for connecting the peripherals. Compared to that, RasPi 1 model B+ comes with a total of 17 dedicated GPIO pins. As described earlier, GPIO works on 3.3V logic levels. 0V to the GPIO means logic zero, and 3.3V to the GPIO means logic one.

Pin number 8 and 10 work for UART communication ports. While connecting these pins to a UART-enabled peripheral device along with UART connection, a ground connection also needs to be shared, as advised in the previous section of UART. To use UART (also called a serial), there is a need to configure the boot file (`cmdline.txt`) located in the RasPi boot partition. We will do it when needed.

Pins 19, 21, 23, 24, and 26 are provided to connect the SPI-enabled peripheral. While 24 and 26 are the CS pins for SPI, the rest of the pins are understandable, as you have read the previous section on SPI communication. If you feel that you don't need SPI communication at a particular moment, you can use it as a GPIO, and unlike I2C, these pins are not internally pulled up. Obviously, why would they be?

Pins 27 and 28 are used to interface the I2C ID EEPROM interface using an I2C connection. Do not use these pins to the peripheral other than I2C ID EEPROM. The preprogrammed EEPROM connected to this port can be looked up while booting time and automatic setup of GPIOs can be done. As it not going to be used in our projects, we need not concentrate on this.

Whether you have the RasPi 2 model B or RasPi 1 model B+, you know all the GPIOs now — really! Let's have some practice on the GPIOs and glow some LEDs to see these GPIOs work. This is the point where you are moving from technical theory to hands-on applications. Kudos!

It's time to glow LEDs!

Let's gather some components and wires first. You need a standard LED (one piece, forward voltage, and 3.3V), wires (one red and one black, with a 2.54 mm female-to-female jumper wire connector), an Ethernet cable, your PC, and the RasPi+.

The standard setup that we had all the time and we will follow is PC (Windows/Mac/Linux) with PuTTY or the terminal installed. There is also the Ethernet connection of the PC with the RasPi, with the entire configuration and setup explained in *Chapter 1, Meeting Your Buddy – the Raspberry Pi*. What I assume now is that your PC is running a live session with the RasPi. Take the LED in your hand and carefully observe that among the two terminals, one terminal is longer than the other; this is the positive (anode) terminal of the LED. The shorter terminal on LED is negative (ground) and it should be connected to pin 6 (GND) of the RasPi. Carefully connect the positive terminal to pin 11 (BCM GPIO 17, refer to GPIO table). Using the wire, you can place the series resistor (330 Ω) between positive terminal of the LED and the pin of the RasPi.

The following diagram is a representation of the circuitry to be made. The dark dots in it represent the connections between the components, wires, and the RasPi:

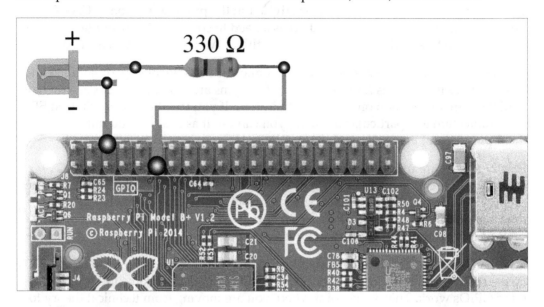

It's time to test now. Insert the SD card and power adapter to start working on the RasPi. You have already installed the wiringPi library in *Chapter 1, Meeting Your Buddy – the Raspberry Pi*. This library comes with a function called gpio that gives you the functionality to access the GPIO directly from the terminal, without even writing any code. Well, codes are better, but for testing the LED, we need not write 5 to 10 lines code when we can do it in a single line. Type gpio -g write 17 1 to glow the LED, or write gpio -g write 17 0 to turn off the LED. Simple, isn't it? These commands can be directly written in PuTTY or in the terminals of Linux and Mac OS X systems. By performing SSH, we can use shell scripting, Python, or C to automate these commands in a single file. We will see all of these in the following sections.

Shell script and GPIO

We already know that shell scripting is a way to execute the commands of a Linux terminal through a single script file. Let's create a script to toggle the LED and generate a pattern on it. What I like is that my LED turns on for a second, then turns off for half a second, and repeats this indefinitely until I press *Ctrl + C*. Open a nano editor using the sudo nano ledblink.sh command, and type the following code:

```
#!/bin/sh
echo "Hi, This is how I glow using Shell!"
```

```
echo "Press ctrl C to exit"
while : ; do
    sudo gpio -g write 17 1
    sleep 1
    sudo gpio -g write 17 0
    sleep 0.5
done
```

The `echo` command does the displaying job, while `:` is used to create an infinite loop. The semicolon (`;`) is the notation used in shell scripting to write two commands in a single line. The `do` command starts the `while` loop, and we can start writing our script to toggle the LEDs inside it. The number that follows a `sleep` command provides you with a delay in seconds. Press *Ctrl + X* and then *Y* to save the changes in your script. In the command line, give execution access to our code, which can be done using the `chmod +x ledblink.sh` command. This command changes the mode and gives the permission to the user to execute the program. You might be remembering that shell script doesn't require compilation, so just type `./ledblink.sh`, and congratulations! You've toggled the LED with the RasPi. Great!

LED blink and Python

Python's script is somewhat lengthy, but it is still easy to implement the GPIO toggling. We will create the same functionality that we created in a shell script, but this time, just for 10 iterations instead of an infinite loop. Open a nano text editor, type `sudo nano ledblink.py`, and then begin writing the following code. Just observe the code and enjoy the blinking:

```
import RPi.GPIO as GPIO
from time import sleep
GPIO.setmode(GPIO.BCM)
GPIO.setup(17, GPIO.OUT)
Print "All Set in Python! Let's Blink"
for i in range(1,10):
    GPIO.output(17,GPIO.HIGH)
    sleep(1)
    GPIO.output(17,GPIO.LOW)
    sleep(0.5)
GPIO.cleanup()
```

Compile and execute the code by typing `python ledblink.py` in the command line. Use `sudo python ledblink.py` if it needs the superuser privileges. Importing the `RPi.GPIO` library is very essential, as it imports the GPIO functionality to the Python scripts and tells Python that the GPIO functions used are for controlling the pins of the RasPi. There is a library named `time` in Python that provides delay, clock, and other time-related functions; thus we only import the `sleep()` function from `time` library as per the needs. We need to set the mode in which we are working. We are following BCM names, so we need to tell the code that whatever the pin number we mention in the functions is in the BCM mode. Then the next function is used to set the direction of the pin, whether it will be used to take the input or output. It sets pin 11 (BCM GPIO 17, refer to GPIO table) to the output pin where our LED is connected. The `for` loop sets the range from `1` to `10`. Unlike an infinite loop, it will run 10 times and then the control will come out of the loop. Inside a `for` loop, the code is self-explanatory; it toggles the LED connected to pin 11 of the RasPi port. In the end, the `GPIO.cleanup()` function will reset the state of the GPIO on exit from the code. It is important to take care of indentation in Python; you can see that there can be a big difference if you change the indentation. For example, change the indentation of `sleep(0.5)` in line with `GPIO.cleanup()` by shifting it to left. You will see a significant change in blinking of the LED; that's what Python is! Takes care a lot on user readability.

Let's blink the LED with C code

Finally we will write the same code in the C language. It's pretty impressive that the libraries available for the RasPi and C are really simple to use. We have already installed the `wiringPi` library to access the GPIO of the RasPi. Let's try the C code now. Edit the `ledblink.c` file in a nano text editor by typing `sudo nano ledblink.c`:

```c
#include<stdio.h>
#include<wiringPi.h>
#include <time.h>
void main()
{
wiringPiSetup();
pinMode (17,OUTPUT);
printf("\n Hi, I am using wiringPi, Blink Blink!\n");
printf("\n Press ctrl C to exit");
for(;;)
{
```

```
        digitalWrite(17,1);
        delay(1000);
        digitalWrite(17,0);
        delay(500);
    }
}
```

The `wiringPi` library should be included with the function call of `wiringpisetup()`. This brings all of the functionality of the GPIO in your C code. Almost every function is self-explanatory, as you have seen in the codes of shell and Python.

You can compile the C code by typing `gcc -o ledblink ledblink.c -lwiringPi`, and execute it by typing the `./ledblink` command.

Press *Ctrl + C* to halt the application code. You can give your LED a breather now; it's been blinking for a long time!

Summary

In this chapter, you understood the electronics fundamentals that are really going to help you go ahead with a bit more complex projects. It was not all about electronics, but about all the essential concepts that are needed to build the projects that we will go through in the next chapters. After covering the concepts of electronics, we took a dive into the communication protocols; it was interesting to know how the electronic devices work together. You learned that just as humans talk to each other in a common language, they also talk to each other using a common protocol.

After covering pin assignments on the RasPi expansion connector, we made an LED blink by programming a GPIO port in the shell, Python, and C languages, which can be useful for showing any kind of indication and decisions made by the RasPi while using the sensors. It was exciting to learn electronics, the communication protocols, and the blinking of the LEDs in your own way. You are now ready with almost all of the information and the skills required to build the sensor-based projects.

In the next chapter, we will do a hack with ultrasonic sensors to measure distance electronically. You will get to know the distance of the ceiling right from your table. We will also build a project as an aid for a visually impaired person to show warnings if obstacles are near. This is really going to be the fun, so let's enjoy together!

3
Measuring Distance Using Ultrasonic Sensors

We, humans have five senses. They are touch, smell, sight, hearing, and taste. However, computers and robots can have as many senses as we want. We can sense different things around us; for example, changes in the temperature can be felt by our skin, but we cannot precisely say what the actual temperature value that we are feeling is. Computers such as RasPi can be used to sense and monitor the surrounding entities. It does the job well, precisely, and untiringly. The computing and interfacing capability of RasPi allows us to interface sensors with it.

Measuring distance using meter tapes and odometers is impractical or inconvenient for some of the applications. If you want to measure the depth of the ocean, how can you use meter tapes? The best option is to use technology such as Sonar or satellites. However, in our homes, labs, and even in our daily life, we often use different ultrasonic sensors for various applications. These applications include overhead tank water-level observation and automated path-finding robots. They also act as an aid to the visually impaired person, as a vehicle parking assistant, and so on.

In this chapter, you will learn the basics of a widely used distance meter, the ultrasonic sensor. You will understand how to use the distance calculation formulas on the RasPi. Further on you will also learn about the hardware setup of the board and also the fundamental requirements to make certain connections. Software level understanding is essential while using the ultrasonic sensor to run the codes and measure the distance precisely.

By the end of this chapter, we will have a project ready. This project will assist the visually impaired to avoid obstacles

We will cover the following topics in this chapter:

- The ultrasonic sensor
- Distance calculation methods
- Hardware setup of the board
- Software understanding
- Installing and running the codes
- Schedule your code while booting up
- Troubleshooting
- Obstacle avoidance system for the visually impaired and blind people

Let's understand the chapter to develop a project for a good cause!

The mysterious ultrasonic sensor

When you go on an expedition to mountains, you must have experienced the echo phenomenon while shouting loudly towards high mountains. You can even experience this phenomenon in a hall that doesn't have interiors such as curtains and furniture (in a new house). The ultrasonic sensor works on a similar principal. Ultrasonic sensors generate ultrasound waves that are targeted towards an obstacle after which they wait for the echo to be heard. However, why don't you hear any sound when you use an ultrasonic sensor? The answer is pretty simple: this sensor works at an ultrasonic frequency, which is higher than the audible frequency range of humans. The human's average theoretical audible frequency range is 20 Hz to 20 KHz. The ultrasonic sensor transmits the sound waves (also called as a sonic burst) higher than 20 KHz frequency. Ultrasonic waves are mainly used because they are not audible to the human ear and also because they provide precise distance measurement over short distances. You can definitely use acoustic sound for this purpose, but it's not nice to have a noisy robot shouting every few seconds. Elaborately, ultrasonic sensors produce sonic bursts and calculate the echo. This echo is received back by the same sensor, calculating the time interval between the transition of the signal and reception of the echo to decide the distance to a target. The concept behind this sensor is almost the same concept used in Radar. This is even more precise than an ultrasonic sensor and works on a very high frequency range (VHF). We can see its overall construction in the following image. This is the representation of the HC-SR04 sensor, which will be used for this project. You can purchase this sensor from any leading e-store or a hobby electronics store near you. The sensor has two cylinders on the board, and these cylinders have metallic nets on top of them. Usually, these cylinders are made up of steel or any equivalent material.

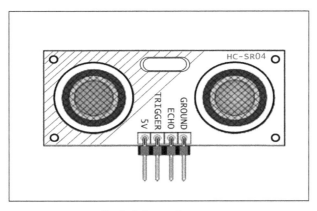

Typical ultrasonic sensor

The sensor shown here has one transmitter and one receiver. For more accuracy, there can be multiple transmitters and receivers. However, this sensor can provide accuracy near ±3 cm within the range of 400 cm. For example, if the measured distance is 270 cm, the actual distance can be 273 cm or 267 cm. Under the cylinders, the sensor has a control circuit that takes care of everything, including the communication with RasPi. There are four pins that come out of the sensor: ground, echo, trigger, and supply. The ground and 5 V supply can be connected to RasPi pins directly. When we give an input from RasPi to the trigger pin of the sensor, the transmitter emits the sound pulses. These sound pulses bounce back from the solid object or surface, and we get the pulse from the echo pin. Then, we calculate the time of arrival of the echo, and we can calculate the distance. There is some physics behind this calculation, which you will learn in the next subsection. This will help us build the code to interface this sensor with RasPi easily. Let's take a peep into the physics behind the sensor.

Distance calculation

Do you know what the speed of sound is? Well, this really depends on which medium the sound wave is travelling in and the ambient temperature as well as elevation from sea level. Brilliant physicists have calculated the speed of sound at the sea level and have found it to be 34,300 cm/s. If you measure the distance under water then the speed of sound is 1,48,200 cm/s. See, it changed drastically when the medium changed; isn't this interesting? This again depends on the water's temperature and so many other entities. While making a project, make sure that you use the correct speed of sound. Here, we are using air as the medium.

We know that,

$$Speed = \frac{Distance}{Time}$$

When we measure the time (the duration of sending a sound pulse and receiving it back), it is measured based on the time taken in going towards the target and returning to the source of the sound waves. However, we want to calculate the time just for the one-way journey in order to measure the distance. For example, we are measuring the distance from point A to B. The sensor will generate the sound from point A. Let's suppose that this sound reaches point B in time T1. At point B, the sound is reflected and reaches back to the sensor at point A in time T2. So, the actual time we measure at the ultrasonic sensor is T = T1 + T2. That is why we need to divide the measured time by the factor of 2.

So now, our equation is as follows:

$$Speed = \frac{Distance}{Time \, / \, 2}$$

We know that the speed of sound is 34,300 cm/s:

$$34300 = \frac{2 * Distance}{Time}$$

Let's simplify it further to use in the code:

$$Distance = 17150 * Time$$

That's all! We know the equation for the distance, and we also know the working principal for the ultrasonic sensor. We will use this exact equation directly in the Python script. So what are we waiting for? Let's start building the project.

Building the project!

It's now time to connect the ultrasonic sensor with the RasPi board. The ultrasonic sensor works on a 5V power supply. Fortunately, we have the 5V supply pin on the RasPi board. We can provide the 5V supply from RasPi to the ultrasonic sensor. However, in reply, the ultrasonic sensor generates a 5V echo signal as an output to RasPi.

It is always recommended that you connect the ground terminals of the devices first and then the voltage supply terminals. This should be followed with almost all the electronic devices we connect with development boards such as RasPi.

As you have read in *Chapter 2, Meeting the World of Electronics*, we know that our RasPi needs the 3.3V level on the GPIO pins to operate safely. So, how do we connect them? It is a serious matter. In this regard, Kirchhoff will help us. With the help of Kirchhoff's current and voltage laws, we can divide the voltage into two parts. If we divide the 5V supply into 3.3V and 1.7V, we can use the echo pulse coming out of the ultrasonic sensor to connect it to RasPi board's GPIO. To divide the voltage, we will simply follow the most common voltage divider circuits used in electronics.

 Avoid connecting the echo pulse pin directly to the RasPi board. This can spoil the board permanently. Always use voltage divider.

Voltage divider is nothing but a combination of resistors with the right values connected in such a style that it divides the voltage. The standard value resistors should be used while creating the voltage dividers so that we can easily get them from the market.

There is a simple theory that we will follow. As shown in the following figure, a voltage divider consists of two resistors (sometimes, only one variable resistor). Usually, the resistor near the input voltage supply (Vin) is called R1 and the one near to the ground is called R2.

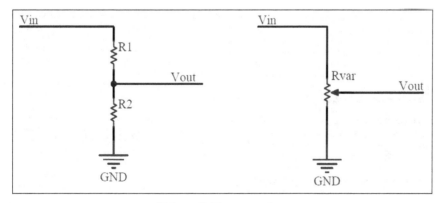

Voltage divider connections

With this configuration, the voltage drop across the resistor R2 should be calculated using this formula:

$$Vout = \frac{R2}{R1 + R2} * Vin$$

Similarly, if you use the variable resistor, also called as a potentiometer (POT), there is a slider in the center. This slider needs to be changed, and we have to look for the desired voltage on a digital voltmeter (multimeter).

I have a 1KΩ resistor with me, and it is one of the standard resistor values available. We already know that we need to convert 5V to 3.3V. So here, Vin will be 5V, and Vout will be 3.3V. Let's calculate the R2 value required to create this voltage divider.

Let's put values into the equation:

$$3.3 = \frac{R2}{1K + R2} * 5$$

Then, let's solve the equation:

$$3.3K + (3.3 * R2) = (5 * R2)$$

We get R2 = 1941.1Ω, and this value of resistor is not available in the market. However, fortunately, we have a resistor of 2KΩ that has a value very near to 1941Ω. The resistors we get from the market have tolerances of 5 percent to 10 percent. This means that resistance with these tolerance values are never exactly same as the tagged value. It always has some variance. As per the calculations, we need two resistors to build the voltage divider from 5V to 3.3V, which are 1KΩ and 2KΩ.

Hardware setup

Connect the devices as per the guidelines given here and enjoy the coding in the next section. You will need these things on a neat wooden table or a table with an insulation sheet:

- An HC-SR04 ultrasonic range finder
- 1KΩ and 2KΩ resistors (if these seller asks about the wattage of a resistor, ask for 1/4 watts, also termed as quarter watts)

If you do not have 1KΩ or 2KΩ resistor, you can use 330Ω and 470Ω resistors, respectively. You can cross-check the calculations done previously. The wattage of resistor states the maximum amount of power dissipated through a resistor. Higher the wattage, higher is the current that can pass through at a certain amount of voltage. P = V * I. So, 0.25W = 5V * I yields, I = 500 mA. Therefore, 1/4 watts can be sufficient on boards such as RasPi.

- One multimeter (if you are a geek and want to measure everything!)

- Female-to-male jumper wires and female-to-female jumper wires

- Breadboard (not compulsory if you can build this by just twisting the wires, but I recommend that you use it for proper connections without shorting each other)

- The RasPi with power adaptor

- An Ethernet cable

- A personal computer

Now, we are ready to build the circuit. The circuit will look like the following diagram if you build it correctly. This diagram is also called as schematics, which represents every terminal of the components and their symbols.

For RasPi, we used the connector in the center with all its 26 pins (for RasPi 1 model B). We already know that RasPi 1 Model B, Model B+, and RasPi 2 Model B have the same functionality on the first 26 pins on header. Let's connect it step by step. It's more enjoyable if you understand the connections and connect it by yourself on the breadboard.

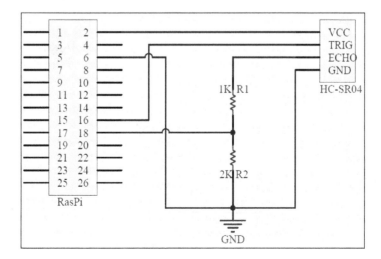

Keep the RasPi powered off while making the connections. You will need the computer after connecting the sensor with the RasPi. Be careful while making all the connections.

However, I have shown the connections on the breadboard in the following figure, which makes the connections easy to understand. The connections are also described here in a step by step. Black dots on the breadboard in the figure show the physical connections of the wire onto the breadboard. We will perform the following steps:

1. Take the breadboard.

2. Connect the sensor on the breadboard. Keep in mind that if you connect it flipped, the connections suggested here would not work. Cross-check the connections, and make sure that you make a correct circuit.

3. Connect pin 6 of RasPi to the ground rail of the breadboard.

4. Connect RasPi's pin 2 to the sensor's 5V pin.

5. The trigger pin can be connected directly with the sensor. In fact, the RasPi sends 3.3V signal to this trigger pin, which is acceptable by the sensor.

6. Take out a connection from RasPi pin 18 to the breadboard, and connect it to the terminal rails of breadboard. From the same row, you can connect one terminal of 2KΩ resistor. The second terminal goes to the same ground strip where RasPi's ground and sensor's ground are connected, as shown in the following image:

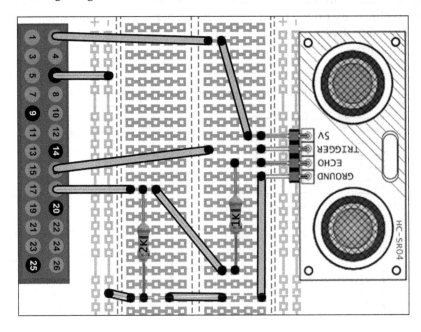

So far, the connections we have done are good to go. Cross-check the connection twice before you power up the RasPi. We have done the procedure on the hardware. We need to write the software (code) to tell the RasPi that we have connected this sensor to these pins and to calculate the distance based on the formulas we derived.

Software setup

Let's start writing the code in Python. This code can be separated into the modules. These modules can then be integrated as a single code to get the distance. Let's summarize the steps here:

- Initial configuration
- Setting the GPIO pins on the default mode
- Sending the trigger signal and listening to the echo
- Calculation of time and distance

Let's see each of them in detail.

Initial configuration

In the initial setup, we need to include or call the GPIO and `time` libraries, which we installed while reading *Chapter 1, Meeting Your Buddy – the Raspberry Pi*, into the Python environment, as shown in the following code:

```
import RPi.GPIO as GPIO
import time
GPIO.setmode(GPIO.BCM)
print "Measuring Distance"
```

As described earlier, the `time` library brings the timing and delay functionalities to the Python code. We will use the BCM mode to call the GPIOs of RasPi, as we did in *Chapter 2, Meeting the World of Electronics*.

Setting the GPIO pins on the default mode

We have connected the sensor's trigger and echo pins to pins 16 and 18 of RasPi, which are GPIO 23 and GPIO 24 in the BCM mode, respectively. We know that the trigger pin is the output pin for RasPi and input pin for sensor. The echo pin is the input pin for the RasPi. Here, it gets the response from the sensor after sending the trigger pulse. The following code depicts this:

```
GPIO.setup(23,GPIO.OUT)
GPIO.setup(24,GPIO.IN)
```

```
GPIO.output(23, False)
print "Setting Trigger pin to zero by default"
time.sleep(1)
```

Sending and receiving the pulses

If we analyze the timing diagram of the ultrasonic sensor to measure the distance, we can easily understand how the sequence should be. The following timing diagram shows the sequence. In the previous step, we made the trigger pin zero by default. Then, we provided a 10µs pulse to the trigger pin of the sensor from the RasPi. Once the sensor receives the trigger signal, it sends the 40 KHz pulse / ultrasonic burst from the built-in transmitter (contains eight pulses).

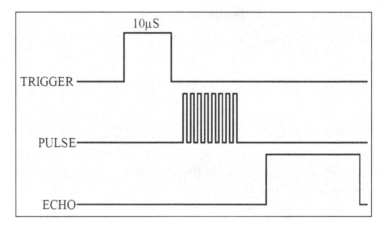

We should expect the echo signal after the ultrasonic burst signal. The length of the echo signal provides us with the timings for the distance calculations. How do we write in a code? The simple way to create a trigger pulse is by turning the pin high for 10µs, as shown in the following code:

```
GPIO.output(23, True)
time.sleep(0.00001)
GPIO.output(23, False)
```

Using this piece of code, we sent a pulse, and the ultrasonic sound burst is generated internally and transmitted into the air. It will reflect and come back to generate the echo pulse. We have to check it continuously until we get the echo pulse. As we saw in the previous figure, the echo pin remains high for a certain period of time. So, we need to calculate the amount of time while the echo signal received stays on. We can use the `while()` loop to continuously monitor the echo pin. We can use the `time` function to capture the time duration of echo while it remains high:

```
while GPIO.input(24) == 0:
    start_time = time.time()
```

This piece of code creates a kind of forever running loop until the echo pin changes its state from 0 (low) to 1 (high). While the echo pin is 0 (or low), we keep updating the timing value (timestamp) in the start_time variable. Once the echo pin goes high, the while() loop breaks, and we get the value of time when the echo pulse went from low to high low state. Next, we need to use the same logic to get the timestamp for the time it remains high. At this instance, we used another variable to store the timestamp so that later on, we can use both the variables to determine the time duration:

```
while GPIO.input(24) == 1:
    end_time = time.time()
```

The while() loop here keeps updating the variable with the latest time while the condition is true.

Calculation of distance

We have the start time and the end time of the echo pulse. So, we have to subtract both the values to get the pulse duration. Once we get the pulse duration, we can then use it for the distance calculation:

```
time = end_time - start_time
```

Initially, in the chapter, we derived a formula for distance calculation, which will be useful here. Let's recall it:

$$Distance = 17150 * Time$$

We considered the two-way journey of sound and then divided the time by the factor of 2.

Converting this formula into the Python script, we get the following code:

```
distance = 17150 * time
print "Measured Distance is :", distance, "cms"
```

We have understood the operation in parts. Let's join them together to get a completely working sensor with the RasPi. Write sudo nano distance.py in the PuTTY command line and start typing the following code:

```
import RPi.GPIO as GPIO
import time
```

```
GPIO.setmode(GPIO.BCM)
GPIO.setwarnings(False)
print "Measuring Distance"
print "Press ctrl+c to stop me"
GPIO.setup(23,GPIO.OUT)
GPIO.setup(24,GPIO.IN)
time.sleep(0.02)
GPIO.output(23, False)
print "Setting Trigger pin to zero by default"
time.sleep(1)
while True:
    GPIO.output(23, True)
    time.sleep(0.00001)
    GPIO.output(23, False)
    while GPIO.input(24) == 0:
        start_time = time.time()
    while GPIO.input(24) == 1:
        end_time = time.time()
    time = end_time - start_time
    distance = 17150 * time
    print "Measured Distance is:", distance, "cms."
```

 You can make use of the inbuilt rounding function in Python to get the rounded values up to three decimal places: print "Measured distance is: %.1f" %distance, "cms."

Press *Ctrl* + *X* followed by *Y* and then press *Enter* to exit and save the code.

Compared to the steps to set up software, there are some minor changes in the compiled code. You will find that we used GPIO.setwarnings(False). This command turns off the warnings generated while compiling the code. When the program is started and the GPIOs are configured, give the RasPi module some time to get ready to take the readings. There is a need to use the while() loop. Using this forever-running loop, we can see the live distance variation by pointing the sensor towards different targets. Before we start to compile the project, check the connections again. Just ensure that the ultrasonic sensor is positioned well and pointed towards the target. Alternatively, just put the sensor perpendicular to the table so that the cylinders remain parallel to the table surface.

And here we go! Type `sudo python distance.py` and press *Enter*. Wonderful, isn't it? As we programmed and executed it, it now shows the output as `Measured Distance is: 9.1321cms`. We have created the ultrasonic distance meter. You can point it at various objects to measure the distance. I have tried to measure the distance from my table to the kitchen to know how much distance I walk every time I go to drink some water. You can stop the code by pressing *Ctrl + C*. This kills any script currently running in the Linux shell. You can use this command if the program is stuck in an infinite loop. This may mean that there is a software or a hardware bug.

> If you are unable to execute the code and the RasPi is having issues with the user privileges, try `chmod +x distance.py` before executing the `sudo python distance.py` command.

Fixing common problems

While building the project, many of us will not get the output the first time. Let's solve the common problems collaboratively. There are some frequently occurring problems while building the projects with the RasPi and the ultrasonic sensors.

Is it showing the distance incorrectly?

You are so enthusiastic to write the code by yourselves, and you might have been mistaken. Check the timings of turning the trigger pin for the perfect amount of time, and check whether you have read the echo pin correctly.

Some of the device response may be slow. I recommend that you add a delay of 60 ms after making the trigger pin low, as shown in the following code:

```
GPIO.output(23, False)
time.sleep(0.06)
```

If you are still not getting the result, try playing around with these values of delay. Also, check for the correct indentation of the code, which might be the major reason that you are facing the problem.

There is a high chance that you may be having so many things on your table that the reflections are very high, as this sensor has a wide angle of sensitivity. If the sensor is touching any object or metallic surface, then in this case also there is a chance to get a false reading.

Is the module not responding?

There can be many reasons when the module doesn't respond. Either you are not powering the RasPi with the adequate amount of current, or the connections are not proper on the breadboard. The sensor can take quite a high amount of current. You might be in need to connect the 5V, 1A charger or adaptor to your RasPi. Check whether the resistors connected are of correct values either using the resistor color-coding method or using the multimeter. Take the multimeter and change the delay value in program to 5 seconds after turning on the trigger pulse. Check whether pin 16 is getting high or not. RasPi's pin 16 should show 3.3V reading on the multimeter.

Are you measuring the distance less than 2 cm?

The sensor will not respond if any obstacle is very close to it. It is very hard for the sensor to measure a very small distance.

A wearable device for the visually impaired

The ultrasonic sensor can provide an added sense to a visually impaired or blind person. You can contribute to the society by making this kind of a wearable device, which can either play a warning sound in the person's ear or a vibration alert on the stick. There are such sticks available in the market, and you can build your own, too.

Your aim is to build a wearable device based on the RasPi and ultrasonic sensor. Also, you will play the warning sound on the headphones that are connected with the RasPi when the visually impaired person is approaching an obstacle that is about 100 cms away. Sounds cool, doesn't it? This project will work for both RasPi 2 Model B and RasPi 1 Model B and B+ users. We will learn this step-by-step process to assemble this project on the hardware and also understand how to code it. Let's dig deeper and understand how this can be done.

Building the hardware

When you run the RasPi, you must have noticed that your project will require the electrical plug socket where we connect the adaptor to power up the RasPi. Our project needs to be wearable. So, as you have rightly guessed, we need a battery. The battery specifications can be quite variable, but you need to stick to 4000mAH–10000mAH and 1A–1.5A USB battery pack. You will easily get the USB battery pack for the RasPi in the online stores. The power banks for mobile phones are also OK, but stick with the current ratings. We also need headphones that have a standard 3.5 mm connector to play the warning tones. Here is a list of the hardware we need:

- A HC-SR04 ultrasonic range finder
- 1KΩ, 2KΩ, and 270Ω resistors
- A 2 mm or 3 mm LED
- Female-to-male jumper wires and female-to-female jumper wires
- Breadboard
- The RasPi
- USB Power Bank 4000mAH–10000mAH with 1A–1.5A current rating
- Headphones with a 3.5 mm connector
- Small wire strips to keep the wires folded
- An Ethernet cable
- A personal computer

Once you have made your testing project in the previous section, additionally, you would need to buy a battery bank and an LED. Any mobile phone's headphones would work fine with the RasPi. Building this device is easy. Connect the ultrasonic sensor with the same breadboard configuration, as we did in the previous section. Additionally, we also connect an LED to the RasPi pin number 11 (BCM mode GPIO 17), with the same method that we followed in *Chapter 2, Meeting the World of Electronics*. The power bank will be connected through the micro-USB jack of the RasPi, where we usually connect our power adapter. Headphones need to be connected to the 3.5 mm jack of the RasPi. The USB cable and headphone wires will be lengthy, so I would like to suggest that you use small wire strips. Small wire strips will help you tie wires and allow you to make your project look neat with tangle-free wires.

This wearable device can be worn at the level of the chest or at the level of the head. You can make use of a cap to attach this device. Pack it beautifully once you finish the programming and testing. Let's have a look at the programming side of it.

Software setup

The device should play a buzz or a beep sound on the headphones. You can get amazing sounds from the www.freesound.org/browse/tags/beep/ web page and download them to the RasPi. I suggest that you download the file that has the shortest duration, for example, 1 second or 2 seconds. Otherwise, it would be annoying to hear the beep sound on headphones for 5 seconds. Either you can download it using the Xming server and inbuilt browser (from PuTTY, using the lxsession command) or transfer it from your PC to the RasPi. But wait, how will you transfer a file on the RasPi?

There is a wonderful program available to copy the files from Windows to the RasPi. This program is known as WinSCP (you can download it from www.winscp.net). Give the IP address, login ID, and password of the RasPi in the FTP mode. You can copy the files from Windows to the RasPi by dragging and dropping them. If you find difficulty using WinSCP, the operating tutorials could be found at https://www.siteground.com/tutorials/ssh/ssh_winscp.htm.

We will make a directory called `project` using the `mkdir project` command. We will perform every operation inside this folder. I am assuming that we are currently at `/home/pi/project` address. You can check it once using the `pwd` command.

As we have already downloaded the essential libraries in *Chapter 1*, *Meeting Your Buddy – the Raspberry Pi*, for the support of `wiringPi` in C, we will now download a library and drivers to play music files on our RasPi. Enter the following commands to configure the drivers in your RasPi. These commands hold good for both RasPi 2 B and B+ users. I assume that you have shared an Internet connection to the RasPi as per the guidelines of *Chapter 1*, *Meeting Your Buddy – the Raspberry Pi*. We are running a PuTTY session on RasPi through an Ethernet cable:

```
sudo apt-get update
sudo apt-get upgrade
sudo apt-get install alsa-utils
sudo apt-get install mpg321
```

It will take some time to download, unpack, and install. Once it is installed, just reboot the RasPi and start the session of PuTTY again on your PC. We can now load the drivers for sound, which we just installed by entering these commands:

```
sudo modprobe snd_bcm2835
sudo amixer cset numid=3 1
```

Now, it's time to write the code that measures the distance as well as notifies the person with an LED indication and a short-duration beep sound. Write `sudo nano project1.py` and start typing the following code:

```
import RPi.GPIO as GPIO
import time
GPIO.setmode(GPIO.BCM)
GPIO.setwarnings(FALSE)
print "Measuring Distance"
print "Press ctrl+c to stop me"
GPIO.setup(23,GPIO.OUT)
GPIO.setup(17,GPIO.OUT)
GPIO.setup(24,GPIO.IN)
```

```
time.sleep(0.02)
GPIO.output(23, False)
time.sleep(1)
while True:
    GPIO.output(17, False)
    GPIO.output(23, True)
    time.sleep(0.00001)
    GPIO.output(23, False)
    while GPIO.input(24) == 0:
        start_time = time.time()
    while GPIO.input(24) == 1:
        end_time = time.time()
    time = end_time - start_time
    distance = 17150 * time
    print "Measured Distance is:", distance, "cms."
    if  distance < 100 and x > 30:
        print "Obstacle Detected"
        os.system('mpg321 buzz.mp3 &')
        GPIO.output(17, True)
GPIO.cleanup()
```

Press *Ctrl* + *X*, followed by *Y*, and then press *Enter* to save and exit the nano editor. We need to take execution permission from Linux to run the code. So, we will enter the following command and then execute it:

```
chmod +x project1.py
sudo python project1.py
```

Try with different distances and check whether the warnings are set properly or not. You can put some text lines to easily debug the code. So far, we had connected Windows PC and the RasPi with an Ethernet cable. However, it is hard when you install it on your head as a wearable and connect the Ethernet cable or keyboard and display it to run the program when the RasPi powers up. Is there any function that whenever the RasPi is booted, the code we just created executes automatically? Fortunately, yes! There is a facility in the Linux kernel called as crontab. Crontab allows us to run the desired program whenever the RasPi is booted. Shell scripting will help us here. We will write a small script in the same directory with the / home/pi/project address. Now, we can write the following shell script to execute the project1.py file we created. Open a nano editor by typing the sudo nano project1_startup.sh command:

```
#!/bin/sh
cd /
cd home/pi/project
```

```
sudo python project1.py
cd /
#end of script
```

This file will be executed by the crontab, so we should give execution access to the file using the chmod +x project1_startup.sh command. Now, we just need to check whether the code executes perfectly or not. Type the ./project1_startup.sh command to execute the shell script file and test whether the sound or LED blinks for less than 100 cm distance. Once this is done, we will edit the crontab file with the following code to enter our shell file to be executed:

```
mkdir chronlogs
sudo crontab -e
```

A file with lot of text will be opened. At the end of the file, just enter the following script and press *Ctrl + X* and *Y* and then press *Enter* to save the file:

```
@reboot sh /home/pi/project/project1_startup.sh
>/home/pi/project/cronlogs/cronlog 2>&1
```

 Note that it is a single line, and hence, it should be inserted without pressing *Enter*.

These commands will also log the errors and execution status to the chronlogs folder in the project directory. We are now ready to test the project execution during boot-up. Test it using the sudo reboot command.

With this, we do not need the keyboard, mouse, Ethernet cable, or PC to configure and run the desired code when the RasPi is started up. Just wear the RasPi and connect the battery supply. Your RasPi is ready to measure the distance and give the indications and warning sound whenever the obstacle is less than 100 cms away. If you want to remove the project from the startup, just re-edit the crontab file and remove the last line we added.

Summary

This chapter gave us lot of knowledge about the RasPi and ultrasonic sensor interfacing, and we enjoyed a lot while building the project. You got to understand how ultrasonic sensors work. You understood that the voltage levels must be the same between the RasPi and the sensors. We used a voltage divider to convert 5V to 3.3V for the RasPi. We set up the hardware and software to start executing the project. We got the distance measurement of any target devices in our lab. At least now, we know how far our ceiling is!

We thought to make a project for a good cause, and we created wearable devices using the RasPi and a USB battery pack. Playing a sound on RasPi device to indicate the alerts was a real fun. It was then interesting to know about crontab to start the file execution at the boot-up of the RasPi module without any need of configuration.

In the next chapter, we will play with the temperature-humidity sensor along with the light sensor. The RasPi does not have any analog-to-digital convertors, but we will do some hardware hacks to interface the light sensors. Let's measure the surrounding environmental properties in the next chapter!

4
Monitoring the Atmosphere Using Sensors

It is highly unlikely to find an equipment without sensors nowadays. Most appliances, such as air conditioners, smoke detectors, fire detectors, gas/CO2 sensors, LCD displays, refrigerators, toasters, thermostats, microwave ovens, and geysers installed in our house have sensors integrated in their circuits to measure surrounding atmospheric entities. When we take a look at our surrounding atmosphere, there are so many entities that can be measured, for example, temperature, humidity, vapor, dust, air quality, various gas levels, wind speed, rain, water quality, light (natural and artificial), presence, motion, moisture, and so on. On a broader level, technologies such as sensor networks and the Internet of Things uses "sensor nodes" that measure one or multiple of these entities and send data to the intended computer or user. The best example of this is the "nest" thermostat (www.nest.com/thermostat/). This device is an example of artificial intelligence that controls and maintains the temperature of your room as per your daily habits and then keeps learning from various scenarios that may arise.

Two types of sensors widely used for such applications are temperature and humidity sensors. In this chapter, we will measure ambient temperature, humidity, and light variations. We will create a kind of a sensor node that takes the decision to turn on or off the tube light and fan. Starting the journey from selecting a proper sensor, you will learn about the DHT sensor and LDR sensors. These sensors play a role in measuring the entities that we will see in this chapter. In this chapter, we will:

- Understand the sensor selection process
- Know about temperature and humidity measurement sensor
- Know about the LDR sensor to measure light variations
- Learn codes for both DHT and LDR sensors

- Build the project to control home appliances
- Learn how to interface multiple sensors that have the similar specifications and properties
- Troubleshoot some of the common problems faced

Let's begin the journey by understanding how to select a sensor and enjoy the amazing world of the RasPi.

Sensor selection process

The process of sensor selection, especially when building an environmental monitoring system, is very confusing in the development phase of a product. We may be unsure about which sensor can be used and which cannot. In the industries, the design and development engineers face this problem every time they start building a project. Sensors are always difficult to control while we interface them with the processor or a controller, because they are sluggish compared to processors. Additionally, we need to take care of a lot of incoming values of sensors that come into the processors. For example, in safety critical systems, such as life support systems or baby incubators, signal conditioning circuitry and signal processing units are used to measure the sensor values precisely and take actions (such as controlling the temperature and oxygen levels for newborn babies or patients) accordingly. Therefore, the first step in selecting the sensor is to find out the exact application. Depending on the application, the sensor selection process should be initiated. In the previous chapter, when we interfaced the ultrasonic sensor, you must have observed that there was a complex circuitry and microchips available on the bottom layer of the ultrasonic sensor. That was a conditioning circuitry that made our job easy. Not all the sensors have conditioning circuitry inbuilt. Some of the sensors come in the form of a single chip. These kind of sensors are tricky to interface. Sensor interfacing is always tricky but not impossible to get working right at the first time. Observation says that whenever the sensor manufacturer's price is costly, it is easier to use and precise in measurement. This is because these sensors are packaged in such a way that they work on most common protocols. This is mostly true, but should not be taken as a statement though.

Criticality of an application

The next thing we notice while analyzing an application is its criticality and the environmental conditions where it is going to be used. Would the application be used in an accident-avoidance system or braking system of an aircraft, in households, or just for a hobby project? Our project is inclined towards a hobby project; we would prefer a user-friendly, inexpensive, and effective sensor for our project. Therefore, we can make a trade-off between accuracy and interfacing leniency. In addition to this, the criticality of the application is not severe or hazardous compared to other sensor systems, such as weather sensors in aircrafts.

Selecting a sensor package

The next step in selecting a sensor should be choosing a package. Some of the applications have a tight space to fit a sensor, and some have a huge space. Compare a mobile phone and a washing machine. Both have a large number of sensors, but the application and room to integrate the sensor inside these applications is incomparable. Package should be selected in terms of the space available on the printed circuit board, final integration of device, packaging, and handling of the device. Basically, electronic components or devices are available in two types of packages (mountings): **through-hole devices (THDs)** and **Surface Mount Devices (SMDs)**. We have used THDs such as resistors and LEDs in the previous chapters. If you see carefully, the ultrasonic sensor had SMD devices on the bottom layer. SMD component pins are very near to each other. Therefore, it is difficult and impractical to use SMD devices on a breadboard, and these chips are hard to solder. For hobby projects, whenever we want to use a sensor in the form of a surface mount chip, we need breakout boards. A breakout board simply expands the pins of the SMD component and gives access to the sensor on a usable pin header. THD components or breakout boards should be the priority for a hobby project.

Sensor properties

After selecting an application and package type, we will move to the properties of various sensors available to measure the same entity. If we want to measure the temperature, we would look at the preciseness of the sensor and electrical ratings. This process is lengthy, as there are many ratings and parameters available for different sensors. List down all the qualities of the various sensors in a spreadsheet.

After listing down the qualities of various sensors, we need to shortlist the best sensors among them. There is always a trade-off between the price and quality of the sensor. Choose a sensor that is inexpensive but also meets the threshold of the quality level required. In sophisticated systems, this compromise of quality over price is never done. However, as we are developing a hobby or household project, this is an effective step to be followed.

Purchasing the sensor

To purchase a sensor, there are many options. You can either choose the nearest hobby shop or you can go for online stores. Asian online stores, such as AliExpress, will provide you cheaper options, and special sensors are easily available there. For the quality products, you need to look towards e-markets such as Digi-Key, SparkFun, Adafruit, element14, and so on. Digi-Key has smart filters to sort the components easily out of thousands available. There is always complete technical information about the product, along with its datasheet. Digi-Key provides you with the facility to purchase the components in numbers from 1 to 10,000 plus. To purchase a single or less number of components, you need to choose the Digi-Reel option. Here, Digi-Key provides the customized reel to selecting the number of components we want to purchase. However, at the time of writing this book, it charges around 7$ extra to pack a customized reel. Adafruit is worth visiting as it provides you with lot information related to the product and competitive prices compared to other e-stores. To purchase a single component, it is beneficial to go for Adafruit.

Available sensors

Some sensors are costly but easy to use; some are cheap but tricky to use. To measure the temperature and humidity, there are multiple options available. Widely used sensors are DHT11, DHT22, RHT03, Si7005, and SHT20. DHT22, DHT11, and RHT03 are available in the THD package, whereas Si7005 and SHT20 are available in SMT packages. We can purchase the DHT11 sensor from Adafruit to start building the temperature and humidity measurement project. The basic difference between DHT11 and DHT22 is that DHT22 provides accuracy, better temperature, and humidity ranges. Therefore, DHT22 is costlier than its predecessor, DHT11. These sensors are a little sluggish while getting data into the RasPi. The advantages of the DHT11 sensor are that they are cheap, and an easy-to-use guide is provided by manufacturer. Other sensors listed here can also be considered. Some of these sensors work on I2C or ADC pins, while others work on a proprietary one-wire protocol.

To measure the ambient light variance (luminance) in your room, we will use a CdS cell (photoresistor). We can purchase PDV-P8001 from Advanced Photonix Inc. and 350-00009 from Parallax. We can easily get the CdS photoresistor from Adafruit; this photoresistor is similar to PDV-P8001. It is also available on SparkFun termed as mini photocell. If there are multiple options, choose the photoresistor that has larger dark resistance values (in ranges of several KΩ - MΩ).

Let's plunge into each of the sensors and understand how they work.

InsideDHT – temperature and humidity sensors

DHT is a combined pack of temperature and humidity sensors. These sensors are ideal for hobbyists who just want to do some data logging. DHT11 sensors are slow in terms of retrieving data. As it is a combined pack, it contains a resistive thermal sensor (a thermistor) and capacitive humidity sensor. A thermistor changes its resistance value depending on the changes in temperature. Technically, all resistors act as thermistors because they dissipate heat when the current is passed through them and are responsible for power losses. However, a resistor's characteristic is that when they are heated up from the external source, the resistance values changes a little bit, which is not so effective to be measured. But this is not the case with the thermistor; its resistance values change in terms of several hundred ohms when there is a slight change in the ambient temperature. A DHT11 temperature sensor provides accuracy near to 2 degree Celsius in the range of 0 degree Celsius to 55 degree Celsius and compared to these values. On the other hand, a DHT22 temperature sensor performs well by providing accuracy near to 0.5 degree Celsius in the temperature range of approximately -38 degree Celsius to 120 degree Celsius.

Let's now understand the humidity sensor. The DHT has an integrated humidity sensor that is not so effective for critical-safety applications. There are polymers or metal oxides arranged in such a way that they create capacitive plates inside a sensor. Due to the effect of humidity, the dielectric constant of the capacitor is changed, and this gives us a variance in the values of humidity. These humidity sensors do not guarantee accuracy, but they are suitable for some applications. The accuracy of a DHT11 temperature sensor is 5 percent in the range of 20–80 percent humidity, while a DHT22 temperature sensor provides accuracy of about 3 percent in the range of 5–100 percent humidity.

Look wise, DHT sensors are generally present in a white or blue colored plastic container with a grid or net on the front. They have four terminals on the bottom layer. Inside DHT, there is a small analog-to-digital conversion IC that provides us with the digital values in two parts: temperature and humidity. Initially, it looks difficult when we try to understand DHT's one-wire (or a single-wire) protocol. With some tweaks, it is simple to use, and the timing must be precise in order to achieve the perfect values from the sensor. Once the timing diagram is understood, we can retrieve data from the sensor easily.

DHT works on single-wire bidirectional bus communication. This means that the data transfer between the RasPi and the sensor works on a single wire with particular timing or delay instructions sent by the RasPi and receives it as a bit stream. Whenever the protocol is not defined, we retrieve the data from the sensor using the bit banging technique. In this technique, the master device sends the exact duration of the pulse (as long as it is recommended by the sensor reference manuals or datasheets) as a request from the master (in our case, the RasPi), and the sensor then responds to be ready. After the predefined amount of time, it starts sending the data, bit by bit, to the master. It is the master's task to receive it at a single bus sequence when the slave has sent it. Any kind of acceptance of chaotic sequence will turn into corrupted data. This means that the sensor (slave) has sent the data perfectly, but the master did not receive it correctly, just because there is no timing synchronization between the master and slave.

Introducing the photoresistor (photocell)

Did you know that the RasPi does not have an analog-to-digital (A2D) convertor integrated inside? This is the biggest drawback while using microprocessor-based development boards such as the RasPi. But do not worry; we have a technique to hack it. A photocell or light-dependent resistor (LDR) is a light-controlled variable resistor. The resistance of an LDR decreases with increasing incident light intensity. More the light, lesser the resistance, and vice versa. The variations in the values vary by about 45 percent, and they shouldn't be used to try to determine precise luminance levels in candela or lux.

Appearance wise, the sensor has a clear shiny, thin surface of glassy material on the top and has two terminals. On the top, there is a photosensitive semiconductor material that is sensitive to light, and we can see the tracks routed in a zigzag pattern.

It's time to introduce you to some mathematics of resistors and capacitors. Understanding the math of resistors is easy, but how are capacitors coming into the picture? Well, there is a phenomenon in the electronics called as a RC timing and RC time constant. Already, the sensor is resistive. We will use the capacitor to utilize this electronics timing constant formula $T = R*C$.

Yes, that's it. The capacitor can be charged through a resistor in T time, where R and C are the values of the resistance and capacitance. We will observe the capacitor to be charged at 50 percent voltage, that is, near to 3.3/2 = 1.6V, using the 1μF capacitor. The RasPi can only read the values near to 1.6V and greater than that. Lesser the light, higher the resistance; lower the charging rate, higher the time to charge the capacitor. For example, if the resistor value is 1KΩ, then the time constant will become 1 millisecond to charge the capacitor.

Due to the series resistance (photocell), the charging of the capacitor becomes slow enough to be read by the RasPi. The sensor values change drastically. The capacitor will make it significantly slow by holding the value for some milliseconds so that we can read the values easily from the RasPi.

Simplifying the scenario, let's take a look at a real-world application to help us better understand the working of the circuit diagram. Imagine a valve (photoresistor) on a pipe (wire) controlled by a lever (light or luminance). This lever helps control an amount of water (voltage) to be stored in a tank (capacitor). The amount of water stored in a tank is immediately taken out by the consumer (the RasPi). The tank here acts as a temporary storage (buffer) to store the water, while the consumer is busy drinking (calculating). This is a good example to understand the working of this hack, isn't it?

This hack can be applied to any resistive sensors, such as thermistor, flex sensors, and pressure and force-sensitive resistors. We can use the same code and hardware explained here to interface these sensors.

Building the project

Whether you have the RasPi 2 model B, RasPi 1 Model B, or Model B+, the hardware setup will remain same and therefore the code too. Once you start coding, you will gradually learn how the RasPi board and DHT11 sensor communicates. In parallel, you will also understand how to interface it to the LDR sensor. You will get to know how multiple sensors can be integrated to get the data. However, first, let's take a look at the circuitry we need to build.

Hardware setup

This project has minimal requirements. It is easy to make the connections on the breadboard, unlike the voltage divider on the previous project. To make this setup, we will require the following devices on our table:

- DHT11 temperature and humidity sensor
- LDR / photoresistor / CdS cellsensor
- 4.7KΩ, 270 Ω, and 10KΩ resistors (if the seller asks about the wattage of the resistor, ask for 1/4 watts; it is also termed as quarter watts)
- 1µF-16V through-hole electrolytic capacitor
- One LED
- One multimeter
- Female-to-male jumper wires and female-to-female jumper wires
- Breadboard
- The RasPi with power adaptor
- An Ethernet cable
- A personal computer

Take a look at the following figure. It represents the connections of our project. The LDR sensor is not a directional electrical component (it operates the same as the resistor). Therefore, it can be connected either way, unlike the LED. One of the terminals of LDR goes to the +5V line of the RasPi pin header, and the second terminal connects to pin 7 on the RasPi pin header (BCM mode GPIO 4). The same terminal that is connected to pin 7 connects to the positive terminal of the 1µF-16V electrolytic capacitor.

 To determine the polarity of the capacitor, you need to check the capacitor closely. There will be a strip marked by - on the body of the capacitor. The strip will have a color in contrast to the color of the capacitor.

The negative terminal of the capacitor connects to the ground (common) pin of the RasPi. The LED can be connected in the same way we usually connect on pin 11 of the RasPi pin header (BCM mode 17):

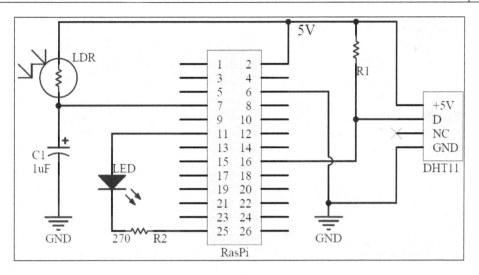

Take the DHT11 sensor in your hand and look at it from the grid side. The following pins from the left-hand side are assigned as VCC (+5V), data, **No Connection (NC)**, and ground (common). Connect the VCC pin of the DHT sensor to the same +5V line where we connected the LDR sensor. The data line of the DHT needs to be pulled up to the +5V line with a 4.7 K or 10 K resistor (R1). Otherwise, there is a chance of getting random data on the RasPi header pin 16 (BCM mode GPIO 23). The ground line will be common for all. Pin number 25 of the RasPi comes under the **Do Not Connect (DNC)** category, but we can use it as a GND pin. All the grounds should be made common either using the ground rail of breadboard or by connecting them to a single ground pin of the RasPi.

Both sensors work well at the +3.3V limit, too. You can experiment it by changing from the +5V wire to the +3.3V pin (header pin 1) of the RasPi GPIO header. Be careful while switching from the +5V to +the 3.3V line. These lines are very near to each other on GPIO header, do not make any short between these two lines. As a result of shorting, the RasPi would be switched off/reset automatically.

Breadboard setup

Once the circuit is understood, it will be very easy to connect components on the breadboard. The DHT sensor's orientation is very important to connect. Follow the image shown here and insert the DHT sensor directly into the breadboard. In the image, the LED connections are not shown as you are smart enough to connect them by yourselves or by taking help of *Chapter 2, Meeting the World of Electronics*. The black dots on the breadboard are the connections to be made either using the male-to-male jumper wires or female-to-male jumper wires:

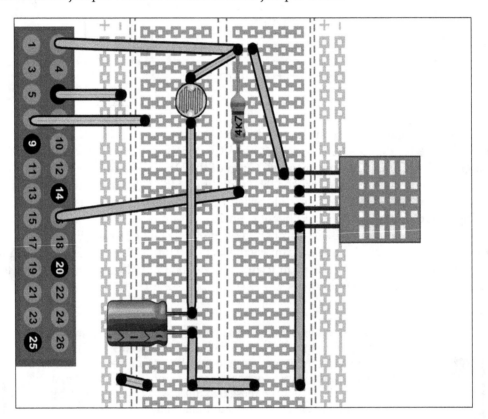

While making the connections, ensure that the RasPi is switched off and carefully verify the connections made. It is good to double-check everything while working with very sensitive hardware such as RasPi, because we do not want to lose our friend and shed tears.

Preparing the code

Working with the DHT11 sensor is quite complicated but not difficult. In parallel with this sensor, we have interfaced the LDR sensor. We have to manage the timings for both the sensors. The problem can be easily solved when we break it down into several instances. We will split the project into smaller steps as follows:

1. Understand the bit banging technique for DHT and prepare the code as per the timing sequence.
2. Write the code for the DHT sensor and test it.
3. Test the code for the LDR sensor.
4. Combine both codes and test it.
5. Add the comparison loop, and make the LED glow to show some indications.

Code the DHT sensor and measure relative humidity and temperature

As we know, the DHT sensor works on single-wire bidirectional protocol. Thus, the synchronization between the sensor and the RasPi is very important. We have to read the sensor at timed intervals. Once the RasPi sends a request, the DHT sensor sends the recently read temperature and humidity data simultaneously. As the sensor is slow and from the timing diagram shown in the next figure, it is recommended that you read the sensor or send the request to the sensor every 5 seconds or more to get accurate data.

 If you have already set up the hardware, you need not remove the connections of the LDR sensor while working with the DHT sensor.

Let's first see the timing diagram to be followed, and then, we will go through the responses received from the sensor. If you observe the upcoming timing diagram image from the left-hand side, the data line is pulled up (high) by a resistor of value 4.7KΩ. Therefore, whenever there is no forceful action, such as pulling down, or no data transmission, the line status is pulled high by default. Understanding the diagram in the sequential manner, we will move from left to right.

The signal has two logic levels: VCC(5V) and GND. Pulling the line high (pull-up) is necessary because logic 1 is the default state to be maintained at the sensor's data line. Therefore, we will have a high state of the line when we initiate the communication between the DHT and the RasPi. The following timing diagram is for our understanding purpose only. The frequency and timing are not populated in this figure.

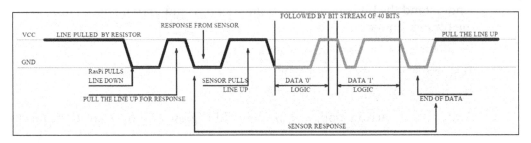

Once the RasPi is ready to receive the data from the DHT, it pulls the line down for more than 19 ms and makes it high again. After some delay, the DHT will sense that the line has been brought down to zero by the RasPi. It then sends the acknowledgement of high pulse of duration 80μS to the RasPi and again brings the line to the low state. Therefore, the RasPi will wait for the 80μS pulse, and then, it will now know that the sensor will send the data. After sending the acknowledgement pulse, DHT11 starts sending 40 bits of data to the RasPi. The data 0 format in the bit stream is nothing but a combination of low level for 50μs and 26–28μs of high logic. Similarly, logic 1 is low level for 50μs followed by high logic level for 70μs.

 Eight bits make one byte. Four bits make one nibble.

The received 40 bits contains 8-bit integer RH data, 8-bit decimal RH data, 8-bit integer T data, 8-bit decimal T data, and 8-bit checksum. The checksum is the total value of all the received data so that we can verify that the received data is correct.

If the data received is 0011 1101 0000 0000 0001 1010 0000 0010 1010 0110, then 0011 1101+0000 0000+0001 1010+0000 0010= 1010 0110 is the calculation.

Adding the first four bytes to verify with the last eight bits, it should be observed that the summation of the first four bytes is equal to the final byte received. In the given example, the received data looks correct. Convert the binary into hex and then into decimal form to get the exact value. We get *Humidity= 0011 1101= 3D (Hex) = 61% RH* and *Temperature= 0001 1010=1A=26 degree Celsius.*

After receiving the data of 40 bits, it's our wish to resend the request to DHT and start the whole process again. DHT keeps sensing the temperature and humidity and keeps the data ready for the RasPi. Whenever it senses a pull down from the RasPi, it initiates the acknowledgement pulse followed by data bits.

Let's get an understanding of the code for DHT11 alone. We will follow the timing diagram and the description given earlier to prepare a code. We will follow the C coding (the `wiringPi` library) for this method. Initially, the data is sent to the DHT11 sensor as a request to send the data. Therefore, the RasPi pin will be made output for a while. Following the next step, the same pin will be made low for the 18ms first. Then again, make the pin high for 40μs and wait for a response from the sensor. To wait for a sensor, we need to observe the data coming from the sensor. Therefore, convert the same pin to the input pin.

Let's take a look at the following code snippet:

```
pinMode(DHT, OUTPUT);
digitalWrite(DHT, LOW);
delay(18);
digitalWrite(DHT, HIGH);
delayMicroseconds(40);
pinMode(DHT, INPUT);
```

The response from the sensor should come within 80μs duration from wherever it is. We need to continuously check for the data now. Now, we can just roughly calculate how many bits would be coming from the sensor and in which duration. Assume that we get all 1s from the sensor. We have considered 1 because it has the longest duration of pulse, that is, 50μs+70μs=120μs. If we consider sequentially reading the method twice (mentioned in datasheet), the total time will be 2*(40bits*120μS) = 96μS. We will wait for the acknowledgement pulse approximately for 100μS. We will run the `for` loop to obtain data from the sensor almost 100 times.

```
for (pulse = 0; pulse<100; pulse++)
{
```

Every operation that follows will be done inside this `for()` loop. We will initialize one value to zero and break the `for()` loop if it reaches the `255` value (setting a timeout).Wait for every bit incoming from a sensor using the `digitalRead()` function in a `while()` loop condition, and whenever the bit state is changed, we will break the `while` loop and save the last state into a variable. This is how it's done:

```
value = 0;
while(digitalRead(DHT) == prev_state)
{
    value++;
```

```
        delayMicroseconds(1);
        if (value == 255)
        break;
}
prev_state = digitalRead(16);
if (value == 255)
break;
```

After studying the timing diagram, we know that the initial pulses are sent to ensure that the RasPi is in the ready state. Therefore, we will wait till the data starts coming in. The first three pulses will be ignored, and we will start reading from the fourth pulse (here, the pulse means a change of state). In addition to this, the change of state should be an even number. Therefore, we will take modulo 2 of the pulse value. Now, here comes some tricky part. The data is coming in a bit-by-bit format. How are we going to store this? We will break the integer format of C into 8 bits. For your knowledge, every integer has a capability to store 8 bits. An array element can be used to store every byte (8 bit). Therefore, each of our bits will be taken as 8 bits, and while storing the data, we will shift to the left (<<) every single bit by one position. We are storing every element of the received data (8-bit integer RH data +8-bit decimal RH data +8-bit integer T data +8-bit decimal T data +8-bit checksum). Therefore, we will place an if loop inside another if loop to change the index of an array, as shown in the following code:

```
        if ((pulse>= 4) && (pulse % 2 == 0))
        {
              data[j / 8] <<= 1;
              if (value> 16)
                    data[j / 8] |= 1;
              j++;
        }
} //FOR LOOP ENDS HERE
```

We also get the checksum in the last eight bits of the data to verify whether the data received is correct. There may be a probability that there might be more data than the actual length. Therefore, to make yourself assured, when we compare the 8-bit value with the checksum, we perform the AND operation with the 11111111 value. The condition to satisfy here is that the sum of all values (initial four bytes) matches with the fifth byte received. If this condition is satisfied, we would print the final humidity and temperature value. Otherwise, we would just discard the data and begin the whole process again after waiting for around 1 second to make sure that the sensor prepares the data to be sent to the RasPi during that time interval.

Each communication process takes about 4 seconds. It is recommended that you send the request for data in average 5 seconds, which means, making the data line low from the RasPi. Adjust the delay accordingly if the data is continuously received as incorrect.

Bit by bit, after storing the data into the array, the data will be printed on the screen, and we should be able to see the relative humidity and temperature values. Following code snippet does the same job:

```
if ((j >= 40) &&(data[4] == ( (data[0] + data[1] + data[2] + data[3])
&0b11111111)) )
{
printf( "Relative Humidity is %d.%d %%and Temperature is %d.%d
'C \n",data[0], data[1], data[2], data[3]);
}
delay(500);
```

We can use an infinite loop, such as while(1) or for(;;), to continuously get the values. However, before that, we have one task pending, which is preparing a code for the LDR sensor.

Code the LDR sensor and measure light variations

The most difficult part of the project has already been discussed. We will now use the LDR sensor to detect light changes using the hack we introduced while studying the RC time constant in an overview of the LDR sensor. Whenever we want to measure or take response from the LDR sensor, we will make the GPIO pin of the RasPi 0 (low) for a moment. Directly, the GPIO low pin makes the capacitor fully discharge to take the readings. Then, after a small duration, we will make it as input pin to check up to what time it charges and reaches up to the voltage level. This should be enough to recognize the RasPi pin as a 1 (high). More technically, when we make the RasPi input pin (GPIO header) as input state, it goes to a high impedance state. Therefore, whenever the voltage reaches 1.6V or above, the RasPi can recognize it as high. Framing the whole sequence of LDR sensor: lesser the light, higher the resistance of LDR. Higher the resistance, higher the voltage drop across the LDR and slower charging of the capacitor. Slower charging of the capacitor, longer the time to reach the voltage level up to 1.6V. The C codes are very fast. Therefore, if we do not limit the readings from the sensor, the RasPi would be flooded with the data. We will use the index (variable index) of 50 to get the data 50 times, and then, we will take the average of the data received:

```
for (index=0;index<50;index++)
{
      pinMode (LDR, OUTPUT);
```

```
        digitalWrite (LDR, LOW);
        delay(16);
        count=0;
        pinMode (LDR, INPUT);
        while (digitalRead(LDR)==LOW)
        count++;
        val[ind]=count;
    }
    sum=0;
    for (index=0;index<50;index++)
    sum+=val[index];
    printf("LDR Value is %d\n",sum/250);
```

The values received by the RasPi are so large that we need to divide it by five times more than the index. According to the hardware configuration, define a pin to the LDR value and make it low for 16ms, as the LDR sensor's resistance values and the RC time constant are changing drastically. After that, set the pin as input (high impedance state) and increment the count value until the pin becomes high. Therefore, whenever there is ample amount of light on the sensor (use mobile phone flash to test), the count value is small and vice versa. Add all the values to the sum variable and take the average of the value. Wasn't this so simple?

Putting all the parts together

We can build a project that understands that the temperature and light are high, and hence, would turn off the home tube lights while automatically turning the fan on! We can integrate the whole project and make an LED glow to show the decision made by the RasPi. Does this sound good?

 Whenever we want to control home appliances, we need to be extremely cautious that they work on 110/230V AC and up to 15A of current. It is not recommended to connect the RasPi to any of the AC mains-operated home appliances directly. Relays (which provide good isolation) should be used to control the appliances. However, it is highly recommended that you perform any task related to relays and controlling appliances under the observation of an experienced electrician.

We can order the relay boards from any e-store. The relay boards are separately powered through an adapter or direct mains supply. The +5V or +3.3V pins of the RasPi should not be shared with these boards. Nevertheless, we can connect the GPIO and ground pins directly to the relay board input/trigger pins. For demonstration/testing purpose, we should connect LEDs on the GPIO port, and once the LEDs are tested successfully, we can interface the relay board. The working will remain the same. For testing, this project we will not require any electrician, as we will make the LEDs glow.

If you decide to use a relay board, do not forget to common the ground pins of both the RasPi and the relay board.

First of all, we will merge the codes of the DHT11 and LDR sensors. To match the timings, some parts of the code will be a new addition to the previous two codes we wrote for DHT and LDR. We will define some `include` libraries, such as standard input/output (`stdio`), standard integer library (`stdlib`) (because we will break the 8-bit integer into binary bits), and `wiringPi` to get support of the RasPi GPIO. We have defined some of the integers as `uint8_t` to define them as 8-bit data.

In code, the LDR, DHT, and LED devices are defined as 7, 4, and 0, respectively. As we are using the `wiringPi` library, they are GPIO numbers that would be same for model B and B+ (previously, it was BCM and Board modes). To know the exact `wiringPi` pin numbers mapped to the RasPi GPIO pin header, type `gpio readall` command in the terminal. This command will give you a table of all the naming conventions followed in the RasPi as output. This is very useful when we code with different libraries.

Open PuTTY on your PC by connecting it to the RasPi through an Ethernet cable and type the `sudo nano dhtldr9.c` command to enter the code to get this project up and running:

```c
#include <wiringPi.h>
#include <stdio.h>
#include <stdlib.h>//Library for integer arithmetic and conversion
#include <stdint.h>//Standard Integer Library
#define DHT 4   //RasPi GPIOHeader Pin 16
#define LDR 7   //RasPi GPIOHeader Pin 7
#define LED 0   //RasPi GPIOHeader Pin 11
int data[5], count, index, k, val[1000],sum=0;
int main(void)
{
    if ( wiringPiSetup() == -1 ) //Check and call wiringPi library
        {exit( 1 );}
    pinMode(LED,OUTPUT);  //Set LED pin to Output
    digitalWrite(LED,LOW);  //Make LED OFF by default
while ( 1 )
{
    uint8_t prev_state= HIGH;
    uint8_t value= 0;
    uint8_t j= 0, pulse;
```

```
        for(k=0;k<5;k++)
            {  data[k] = 0;}  //Clear the data in array

        pinMode( DHT, OUTPUT );
        digitalWrite( DHT, LOW );
        delay( 18 );
        digitalWrite( DHT, HIGH );
        delayMicroseconds( 40 );
        pinMode( DHT, INPUT );  //Initial Handshake between Sensor
and RasPi

        for ( pulse = 0; pulse < 100; pulse++ )
        {
            value = 0;
            while ( digitalRead( DHT ) == prev_state )
            {
                value++;
                delayMicroseconds( 1 );
                if ( value == 255 )
                    {break;}
            }  //while loop ends here
            prev_state = digitalRead( DHT );
            //Catching the data into the data Array

            if ( value == 255 )
                break;
            if ( (pulse >= 4) && (pulse % 2 == 0) )
            {
            data[j / 8] <<= 1;
            if ( value > 16 )
                data[j / 8] |= 1;
            j++;
            }
        }  //for loop ends here

        if ( (j >= 40) &&
    (data[4] == ( (data[0] + data[1] + data[2] + data[3]) &
    0b11111111) ) ) //Verifying the Checksum, Validate and Print the data
        {
            printf( "Humidity = %d.%d %% Temperature = %d.%d 'C \n",
                    data[0], data[1], data[2], data[3]);
        }
        else
        {
```

```
        printf( "Data received is corrupted, what did you do?\n" );
    }
    pinMode(DHT, OUTPUT);
    digitalWrite(DHT, HIGH);

    //LDR code initialization
    for (index=0;index<50;index++)   //Setting Code to get 50 values
and store into array
    {
        pinMode (LDR, OUTPUT);
        digitalWrite (LDR, LOW);
        delay(16);
        count=0;
        pinMode (LDR, INPUT);
        while (digitalRead(LDR)==LOW)
            count++;
        val[index]=count;
    }
    sum=0;
    for (index=0;index<50;index++)
        sum+=val[index];   //Take sum of all values in Array
    printf("LDR Value is %d\n",sum/250);   //Take average of sum and
                        scale by 5 and print the data
    if(data[2]>25 && sum>4000) //Check the Temperature and Light
Condition
    {
        digitalWrite(LED,HIGH);
        printf("TEMPERATURE IS HIGH >> FAN ON, LIGHT IS LOW >>
TUBELIGHT ON\n");
    }

    delay(500);   //Adjusting the Delay to match DHT11's minimum
request time
```

Now, after typing the code into your RasPi, just press *Ctrl + X* followed by *Y*. Then, press *Enter* to save the changes. Enter the `gcc dhtldr9.c -o dhtldr9 -lwiringPi` command to compile the code. Then, enter the `sudo ./dhtldr9` command to execute the code.

> It is very important to execute the code with root privileges in order to change the state of GPIO pins of the RasPi. Therefore, we need to use `sudo` to execute the code.

The following screenshot shows the output of the `sudo ./dhtldr9` command:

```
pi@raspberrypi ~/ch4 $ sudo ./dhtldr9
Humidity = 38.0 % Temperature = 28.0 *C
LDR Value is 8765
TEMPERATURE IS HIGH>>FAN ON, LIGHT IS LOW>>TUBELIGHT ON
Humidity = 38.0 % Temperature = 28.0 *C
LDR Value is 9112
TEMPERATURE IS HIGH>>FAN ON, LIGHT IS LOW>>TUBELIGHT ON
Humidity = 38.0 % Temperature = 28.0 *C
LDR Value is 8686
TEMPERATURE IS HIGH>>FAN ON, LIGHT IS LOW>>TUBELIGHT ON
Data received is corrupted, what did you do?
LDR Value is 8700
```

Troubleshooting common problems

We have covered almost everything on interfacing both of these sensors together. If you are still facing the problem in getting/reading the data, you could go through some of the common problems faced in the following sections. In most of the problems, it is advised that you check the connections made on the breadboard and correct the RasPi GPIO pin.

Received DHT data is not valid

The following points need to be kept in mind when the received DHT data is not valid:

- First and foremost, as mentioned earlier, the received data directly depends on the connections you have made. Carefully check the connections of the pull-up resistor between data line and +5V line.

- There can be a wrong value for the resistor. If you have connected 4.7KΩ, try to change it to 5KΩ, or you can use a maximum 10K resistor. A high value of the resistor can turn data into false or corrupted values.

- Check the DHT orientation. The pin on the left-hand side, seen from front, is the VCC pin.

The LDR sensor gives a zero value

The following points need to be kept in mind when the LDR sensor gives a zero value:

- Check the capacitor value connected between the LDR sensor and the ground. A small value of the capacitor voltage cannot build up much charge. It is highly unlikely that the 1µF capacitor comes less than 16V, but check that it is not lesser than 6.3V rated.

- Try running the Python code for the LDR sensor. If it shows the correct values, then there is no problem with the connection. As mentioned earlier, the Python code is simpler than the C code. Just check it in the following code:

```
import RPi.GPIO as GPIO
import time
GPIO.setmode(GPIO.BCM)
value = 0
GPIO.setup(4, GPIO.OUT)
GPIO.output(4, GPIO.LOW)
time.sleep(0.2)
GPIO.setup(4, GPIO.IN)
while (GPIO.input(4) == GPIO.LOW):
    value = value + 1
print value
```

Is the voltage correct?

The following points need to be kept in mind when checking whether the voltage provided is correct:

- As mentioned earlier, both the sensors can work on either +5V or +3.3V. As we have connected a resistor on the data pin of the DHT sensor, it impressively drops the voltage to a sufficient level to maintain the level of the line. On the other side, the LDR sensor itself is a resistor to drop enough voltage and charge the capacitor. Insufficient voltage cannot be the problem with the LDR sensor, as it can work from 1.8V to 100V.

- Some of the breadboards have discontinued lines of +5V and ground on the side. Check the connectivity of supply lines using the multimeter, and be assured that the ground is properly connected and shared with the RasPi.

Summary

In this chapter, we started with the process of selecting a sensor, in which we got to know that the sensor selection process is not as easy as it looks. There can be so many parameters, and none of them should be neglected while selecting a sensor. This affects the system according to the criticality of the application.

Following the explanation to introduction of the DHT and LDR sensors, you continued to learn with the tricky DHT sensors. The timing diagram was somewhat complex to implement. Due to lack of onboard analog-to-digital convertor, you learned about the RC time constant and use of the capacitor to integrate any resistive sensor with the RasPi and the LDR sensor in particular. Finally, we integrated everything and made a project that can be called a small project of home automation. We were able to integrate the LDR and DHT sensors and make a sensor node using the RasPi. Using the data of temperature and light captured from sensors, we controlled the appliances such as tube, light, and fan.

In the next chapter, you will learn to integrate the analog-to-digital convertor chips that are useful to integrate any sensor with the RasPi. We will understand why it is important to use the ADC chips. ADC chips can enable the RasPi to be a sensor station, and we can log as much data as we want.

5
Using an ADC to Interface any Analog Sensor with the Raspberry Pi

It is known in the RasPi community that the **analog-to-digital convertor** (**ADC**) is not integrated into the RasPi because the onboard processor (BCM2835) does not have any functionality to perform the conversion. Had the Raspberry Pi integrated ADC module, it would have been very easy to interface the sensors, since ADC modules are used to convert analog values into discrete and digital values. Most sensors provide us with the analog signal on their output pins. These analog signals are not understood by the processor as they are varying in nature. Interfacing the ADC module ensures that the correct data will be read by the processor to make appropriate decisions. In a simple manner, the ADC module creates a bridge between the analog sensor and the RasPi core processor. More specifically, ADC modules convert analog signals into digital data and provide the converted data to the RasPi through certain protocols such as I2C, SPI, or UART. Compared to the RasPi, the AVR, Arduino, and PIC18 boards have up to eight analog inputs and internal ADCs. However, the RasPi has a lot of other functionalities to compensate for the absence of an ADC module, such as handling an operating system, full-HD video streaming, camera support, and much more. We can always provide the ADC conversion functionality externally.

Earlier, ADCs were made using discrete electronic components, using BJTs or operational amplifiers and resistors. All the credit goes to integrated circuit developers for integrating all such things and giving them to us in the form of a single chip. ADC modules are used in almost all kinds of smart equipment, and whenever there is a requirement of interfacing a sensor, ADC modules are used in one way or the other — either inside the microcontroller or as a dedicated signal conditioning network to take care of incoming signal from ADC.

In this chapter, you will learn how to use and interface the ADC module. You will initiate your learning from the basics of ADCs and the simple mathematics behind it. Once you've learned the calculations to be done to acquire the analog value, we will take a look at a specific ADC IC, known as MCP3008, manufactured by Microchip Technology. You will then gain the essential knowledge on MCP3008 required to prepare generic hardware and a form of a sensor station that will allow you to connect any sensors with it. Similarly, we can then write the generic software to scan any channel of the ADC convertor chip through the RasPi. Using this software, we will then be able to convert the sensor station into the data logger. This data can finally be saved in the `.csv` file for post-processing.

Let's cover the ADCs first.

Analog-to-digital convertors

In this section, we are going to take a look at the types of ADC modules and the conversion process of the signal. We will get insights into answering these questions: How the ADC module takes care of the incoming data? Do all the ADC modules need external signal conditioning circuitry and amplification? How does the ADC module provide the data output? How is the quality of the ADC measured? What are the calculations required while processing the data?

ADC modules are a part of the signal conditioning circuitry in a microprocessor or a microcontroller. Unlike the RasPi, most microcontroller and microprocessor-based boards have ADC modules embedded, and the signal conditioning circuitry too. Most sensors can be directly integrated with these kinds of processors. Practically, in sophisticated systems, these conditioning networks have many stages for taking care of the incoming data. Some processors have dedicated signal processing cores or units to handle the data. ADC and **digital-to-analog conversion (DAC)** topics are so vast that one could write a whole book on explaining the functionality and projects on it. Without going into the depths of ADC, we will cover the basic terms and functionalities.

Data reception and signal conditioning

When data enters the processor, the ADC module's first task is to filter the unwanted noise from the received sensor signal. This filter is known as the anti-aliasing filter, which is used to avoid the aliasing effect observed during sampling. To meet the requirement of the Nyquist's sampling theorem, the sampling rate should be double (or greater) the signal frequency. Anti-aliasing is an effect in which the signal's frequency band stretches and gets overlapped while sampling the signals, which creates an unwanted noise and crosstalk issues. To overcome this phenomenon, anti-aliasing filters are used to ensure that the signals received are in the desired frequency bands. Sometimes an additional filter, such as a band-pass filter, is designed to filter out undesired frequencies.

The amplitude of the signals coming from the sensors is usually weak and varies a lot. Nowadays, advanced sensors have the capability to provide a signal powerful enough to be fed directly to the microprocessor. However, this is not always the case. In some analog data acquisition systems, we require the amplified signal to the filter input. For that operation, we need the amplification stage prior to or after the filtering stage.

Amplification

Sometimes the signals received may appear similar to noise signals. Some smart ADC modules discard the incoming data due to lack of enough voltage or current (basically low power) of the signal. In this case, we use current or voltage or power amplifiers to boost the current or voltage or power level of a signal respectively. For example, a temperature sensor or motion sensor provides very weak signal voltages (in mV), and this range of voltages is too low for the ADC module to start the conversion process. In this case, it is required to boost the voltage to the required voltage range. In most circuits, **Operational Amplifiers (Op-Amp)** are used for amplification of voltage and currents. The amplifiers designed to implement in signal conditioning are log amplifiers, instrumentation amplifiers, peak detectors, and many more.

Sampling and quantization

Sampling and quantization are the terms that are widely used in electronics and communication theories. These theories are too complex to understand for a beginner in electronics. By definition, sampling refers to the process of acquiring voltage levels (at a predefined frequency) from the continuously incoming data at an analog input pin or at the output of an amplifier. Quantization is the process of mapping or shortening down a large amount of input values of the sampler into a countable set of numbers (in this case, the combination of digital bits). Quantization always produces a quantization error because it rounds off continuous input values and makes approximations to its nearest values.

Types of ADC

With respect to electronic design, there are multiple design structures available for different kind of ADCs. Widely used ADCs include Flash Convertor ADCs, successive approximation ADCs, sigma-delta ADCs, and many more. A flash Convertor ADC uses clock pulses and comparators and operates in parallel mode. It is the fastest way to convert analog data into digital bits. A successive approximation ADC also uses a comparator to shrink the range of input voltages. It makes approximations by comparing the input voltage with the internal DAC and stores the result in a successive approximation register. A sigma-delta ADC samples the signal and filters the desired signal frequency. It converts the signal into analog frequency pulses and counts these pulses at regular intervals, so the pulse count divided by the interval gives an accurate digital representation. ADC types and descriptions are endless and cannot be described in this section. Let's cover something more important now.

Resolution of the ADC

Resolution of the ADC is one of the most important terms when you describe or learn about the ADC. Resolution of the ADC refers to the number of values that can be produced over the range of analog input voltage levels. For example, an 8-bit resolution of an ADC can have 256 different combinations of the values because $2^8 = 256$. In general, the number of voltage intervals can be defined by $2N$ levels, where N is the resolution of ADC in bits. When we talk about resolution, we always take reference voltage into account. Reference voltage is another input to be given through the dedicated input pin of the ADC. It reflects the maximum value that the ADC can measure at the analog input pin. Therefore, most ADCs have two types of input: the reference voltage (mentioned as V_{ref}) and the analog input. Let's take the reference voltage as 5 volts and the resolution of the ADC as 8 bits; then *5V/256 = 19.5mV*. Therefore, the ADC convertor cannot understand voltage changes on the analog pin under 19.5mV.

In other words, the ADC module is no more sensitive than 19.5mV. Compare this situation with 12-bit ADC, which can be calculated as *5V/4096 = 1.2mV*, pretty fair, isn't it? Now we can detect changes in the analog input signal with the sensitivity of 1.2mV.

There are two ways to improve the resolution of an ADC. One simple way to do this is to decrease the reference voltage, and the second way is to choose an ADC module with higher bits of resolution.

Reducing the reference voltage is not recommended in most cases because it will narrow down the window of maximum voltage that can be detected by the ADC. For example, if the reference voltage is 3V, then the ADC cannot detect the changes happening over 3V because it saturates at the reference voltage. It is always recommended to use a higher-bit ADC module instead of reducing the reference voltage. The processors, which have an integrated ADC, do have a programmable reference voltage, which can be set during runtime.

The math behind ADC

As you learned, you know that the principle behind the ADC convertor is to convert the sensing parameter into voltage levels. For example, let's take a temperature sensor and its working parameters into account. A good temperature sensor works on a 5V input supply with 10-bit resolution, 5V reference voltage, and temperature-to-voltage conversion equal to 1mV per degree Celsius. It senses temperature in the range of 0 degree Celsius to 50 degree Celsius. Therefore, at 50 degree Celsius, it will provide the maximum 5V output at the analog pin. The digital conversion equation goes this way:

Digital number = $(2^{resolution} / V_{ref})$ * Vin

Let's calculate the value of the digital number converted from the input voltage read at the analog pin:

10 bits; therefore, $2^{resolution}$ = 1024

V_{ref} = 5V

Suppose ADC has read 2.9V at the ADC input pin. Then Vin = 2.9V yields Digital number = 594 by rounding off.

Now the resolution of the ADC is V_{ref} / 1024 = 5/1024 = 0.00488 V/count, and the temperature-to-voltage conversion scale factor is 1mV per degree Celsius.

Therefore, to convert the obtained digital value in the temperature value, we use this formula:

Temperature = (594*0.004882)/0.001 = 28.9990

> To avoid floating value calculations, use the following generic formula: Temperature = digital read value * 4882. This will give you six- to seven-digit integer. Put the decimal point after first two numbers, for example, 594*4882 = 2899908. After putting the decimal point after the first two numbers, 28.9908 is the value of temperature.

If you carefully see, you'll notice that the final temperature value is just multiplied by 10 of the sensed input voltage provided by the temperature sensor. This happened because we had 10-bit resolution, which made the mapping count range from 0 to 1023. Try it yourself with the 12-bit ADC values and check the precision and the value change.

Data output

Once the conversion is over, the ADC module provides data on its output port. When choosing an ADC module for boards like RasPi, there can be two options: we have to choose an ADC module that provides either parallel digital output or data on a known standard protocol bus such as UART, SPI, or I2C; or a proprietary protocol.

MCP3008 for analog-to-digital conversion

The MCP3008 ADC is based on successive approximation ADC architecture with a resolution of 10 bits. The MCP3008 has eight input channels for interfacing the sensors on the pins. If so many channels are not required, we can select a four-channel input ADC chip known as MCP3004. These chips are the best way to interface sensors because they support the SPI slave mode to be interfaced with the GPIO pins of the RasPi models A+, B, and B+ and RasPi 2 model B. This MCP3008 chip should be purchased in a **dual-in-line** (DIP) package so that we can easily insert it on a breadboard while preparing for the project:

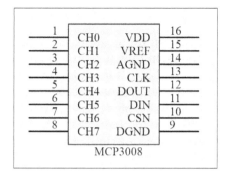

MCP3008

Channels

In the preceding figure, the pin out of MCP3008 is represented. Pins 1 to 8 to the left are CH0 to CH7, which are used for analog input channels 0 to 7 respectively.

 If you have had an enough exposure to electronics in the past, you might have heard about differential pair signals. These differential pair signals are used to eliminate the common-mode noise present in incoming analog lines.

Among these eight channels present in MCP3008, we can use pairs and make it as a four-input ADC with pseudo-differential inputs. The difference between pseudo-differential and fully differential modes is that in the pseudo-differential mode, the ground is separated by a low-value resistor from the common ground for better performance.

The example of a sensor that works on differential modes is ACS726 for current sensing applications in industries for over-current detection and load detection. To avoid complexity in the hardware, we will use single-ended operation, where one channel will be used to take the input from the sensor to one of the channels of MCP3008.

Ground

In MCP3008, there are two grounds available, AGND and DGND. It is best practice in hardware design that the analog ground and digital ground should not be made common or tied together. There should be a filter or ferrite bead to improve the noise immunity of the digital system. In precision devices, this must be followed, but again, our project doesn't need precision or sophisticated filtering. We are happy to tie the analog and digital grounds together, and it is all okay to do that in this project.

SPI

MCP3008 employs four-wire SPI communication. The device is capable of a 200 ksps (kilo-samples per second) conversion rate. It is important to maintain the minimum clock speed while working with SPI-based modules. In the case of MCP3008, the timing between the sampling and data output should be at least 1.2 ms to effectively get the data at the RasPi's SPI port. The pins present on MCP3008 for SPI are CSN, DIN, DOUT, and CLOCK, which are the same as CS, MOSI, MISO, and CLOCK respectively on the RasPi's GPIO port—or similar for any other SPI devices. CSN is used to initiate communication with the peripheral slave device.

CSN is an active low pin. Remember that you have to pull down the signal to zero to start the communication. When it is pulled to high, MCP3008 will end the conversion and the device will go into the low-power standby mode. Between two ADC conversion requests, CSN must be pulled to high.

Reference voltage

As described earlier in the *Resolution of the ADC* section, you learned the importance of the reference voltage. It is the same pin where we need to give desired reference voltage. One interesting thing to know is that there are dedicated chips available for creating precision reference voltage for ADC convertors. For example, our ADC chip is of 10-bit resolution and works on 3.3V. There are some chips available that provide you with precise 3.072V reference voltage. You will definitely think, "why 3.072V?" If you don't remember the formula mentioned in the *Math behind ADC* section, here it is again:

Digital Number = $(2^{resolution} / V_{ref})$ * Vin

Putting our values in the formula, we get this: $2^{10} = 1024$; $V_{ref} = 3.072$; Digital Number = Vin / 3.

It gets so easy to calculate the digital number from the incoming Vin voltage read at the input pin. For our application, we will simply tie the reference pin to the 3.3V so that we can have simplicity at the hardware side.

Supply voltage

It is mandatory to provide stable input voltage for the ADC modules, because if the input supply is noisy or unfiltered, it can directly affect the reference voltage of the ADC. This creates a long chain of errors, as the digital number depends directly on reference voltage. We will provide power supply of 3.3V through the RasPi's expansion port. This voltage is near to stable as it comes from the internal low-dropout voltage regulator.

Making your own sensor station

It is essential to gain knowledge of the ADC modules that we have gone through, and it is good to go with the hardware setup. We know that MCP3008 has eight channels to interface eight different single-ended output analog sensors. What if we build a project that has as many sensors as eight? Do you want to create generic hardware that has the capability of connecting any sensor with your RasPi? This section will contain the development of generic hardware that can seamlessly work with RasPi to interface whichever sensors you want. We can call that hardware by giving a name as a sensor station.

 Not all sensors can be directly interfaced through MCP3008 with the RasPi. As described in the introduction to the ADC convertor, some sensors' output is so noisy or weak that it needs external filters and amplification for those respective sensors. It is recommended to read the datasheets of that particular sensor to know the required additional circuitry to be interfaced.

Until *Chapter 4, Monitoring the Atmosphere Using Sensors*, we used breadboards to create rapid prototyping hardware. We can prepare a circuit on a clean and reusable **general-purpose circuit board** (**GPCB**) by soldering the components. A GPCB can be useful for fulfilling our requirement for making the generic hardware. Still, if you don't want to buy the GPCB, this section will give you the idea of making the circuit on a breadboard. This is because preparing a circuit on a GPCB requires some soldering instruments, materials, and skills of soldering; it would rather be easy to use a breadboard. But there is much more fun in building a hand-soldered project. If you don't have soldering skills, you can build up your skills in an hour or two by practicing on junk hardware. Give it a try!

Here's the list of the hardware we need to purchase in order to start the hardware development:

- A general-purpose circuit board, dual-sided and solderable (2.54 mm pitch)
- Soldering iron (pencil type, 30 W to 50 W) and soldering core with flux
- MCP3008 (eight-channel) or MCP3004 (four-channel and DIP package)
- A single-stranded wire (20-30 AWG)
- A wire stripper
- One dual bergstik connector (male, 2 x 13 pins, 2.54 mm pitch); one single bergstik connector (male, 1 x 8 pins, 2.54 mm pitch); and three single bergstik connectors (male, 1 x 5 pins, 2.54 mm pitch)
- A GPIO ribbon cable for the RasPi 1 model B (26-pin) or B+ (40-pin) and the RasPi 2 model B (40-pin and female to female)
- LM35 or LM36; also known as TMP36 or temperature sensor (TO-92 package)
- Male-to-female and female-to-female jumper wires
- The Raspberry Pi, a power adapter, an Ethernet cable, and a personal computer

Place these components on your workbench and start building the circuitry, as shown in the following schematic diagram:

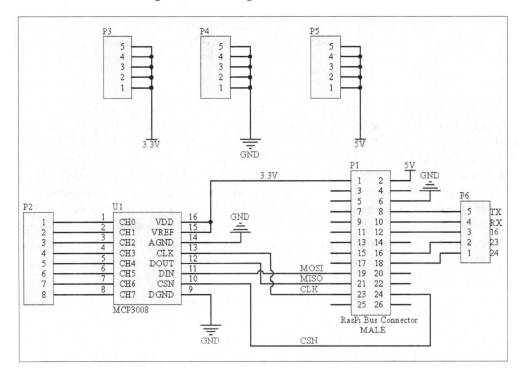

This hardware can now be used to create a generic interface to retrieve data from up to eight different sensors. This hardware setup will be done apart from the RasPi board connections. Once this circuit has been built, we will interface it using the ribbon wire. Let's cover the connectors in the schematic diagram. The second connector from the rightmost area of circuit, labelled as P1, is the 2 x 13 bergstik pin male header of the RasPi 1 model B.

If you have the RasPi 2 model B or the RasPi 1 model B+, then you'll have to use the 2 x 20 bergstik pin male header in place of the P1 connector. The rest of the circuit remains the same.

With the P1 connector, we join the P6 connector in the right to get the functionality of GPIO and UART communication. If needed, we can use it to toggle LEDs or to communicate with some other board or hardware. Pins 19, 21, 23, and 24 are the standard four-wire SPI interface connections net, labelled as MOSI, MISO, CLK, and CSN respectively. On MCP3008, we have provided +3.3V to VREF and VDD of the chip. In our new calculations, we will use 3.3 as V_{ref} while developing the software. The analog and digital ground should be tied together while soldering the circuitry on the GPCB. The P2 connector in the leftmost area of the schematics will be used to connect the analog output pins of different sensors. Connectors P3, P4, and P5 can be used to connect +3.3V, Ground, and +5V to the sensor as per the requirement.

Hold the general-purpose board in your hand and you will notice that it has copper-plated holes. These copper-plated holes can be used to solder the components. Take a look at the top view of the representational circuitry built on the GPCB, as shown in the following diagram:

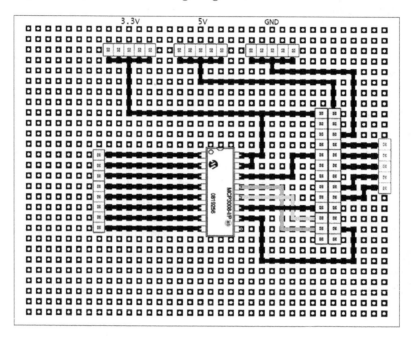

The circuit looks straightforward and same as the schematics. By keeping some space on the left side, insert the connectors into the suggested positions. Here, the 2 x 13 bergstik pin male header can be used to connect the RasPi GPIO using the standard GPIO ribbon cable. If you have model B+ or the RasPi 2 model B, then you can also use the same hardware with the 2 x 20 pin bergstik header. Instead of wiring, you can pour the solder to connect the chip and the connectors. The black lines shown in the preceding diagram are the soldered pads on the GPCB. Just use the iron and solder along the entire path as shown in the figure. One thing that needs to be taken care of is that the MISO, MOSI and CLOCK signals have to be connected using the wires, as they are crossing each other. Therefore, they are represented in grey. Without wiring, it is not possible to solder the whole line from the RasPi connector to MCP3008 without shorting each other.

Soldering irons are too hot when they are at their peak temperature. Be extremely cautious while using them. There could be the chance that you put the iron on wires, your own table, or your own hands! Use a soldering iron stand to place the hot soldering iron when not in use.

While soldering the MCP3008 chip, do not keep the soldering iron near the leads of the IC for a long time. The soldering irons are typically hot, from 250 degree Celsius to 400 degree Celsius. Long exposure to higher temperatures can damage the IC.

After preparing the hardware, we are ready to prepare the code by powering up the RasPi and connecting it with the PC over SSHing through PuTTY using an Ethernet cable. At this moment, there is no need to connect the hardware module that we created just now. Once we write the software as described in the next section, we then will connect hardware module to the RasPi using the ribbon/bus connector.

Generic software preparation

In future, there will be lot more sensors introduced than those you are working with. It will be great if you develop software that is built to use in any of your analog data acquisition projects. Preparing generic software without any errors will reduce developing time in future, and you can rapidly build projects just by adding the working generic software and calling the functions whenever needed.

We know that the MCP3008 is interfaced through the SPI protocol. To use the SPI protocol, we have to install some additional packages on our RasPi. First of all, we need to make our RasPi up to date. Enter these commands to update and upgrade the OS to the latest kernel package. If you haven't performed any update after the fresh installation, then this may take a long time. Once the process is done, restart the RasPi module to perform the normal operations:

```
sudo apt-get update
sudo apt-get upgrade
sudo reboot
```

After this process, we have to make the RasPi board ready with the Python development packages that will bring the default static library packages and header files of the standard Python library:

```
sudo apt-get install python-dev
```

Once this installation is done, we have to enable the SPI functionality for the RasPi board. Because of its rare use, the SPI port is disabled by default. Therefore, it is listed under the `blacklist` configuration file under the `/etc` directory:

```
sudo nano /etc/modprobe.d/raspi-blacklist.conf
```

This command will open the configuration file in a nano editor. Find out the `spi-bcm2708` line by scrolling down with keyboard. Put the # sign before this line to avoid blacklisting of the SPI port during the startup process. Press *Ctrl + X* and then press *Y*, followed by *Enter* to save the configuration file. After this process, we have to reboot the RasPi again:

```
sudo reboot
```

Now, check whether the RasPi Linux kernel has loaded the SPI drivers while booting or not. Enter the following command and check for the SPI drivers in the list:

```
lsmod
```

It is always tricky to work with protocols such as SPI unless some handy library helps us. Thanks to `doceme` for sharing the `python-spidev` library on GitHub, which provides us with the functionalities to pull and push the data from the SPI port. Create a new folder for your project and download the `spidev` library:

```
mkdir mcpgeneric
cd mcpgeneric
wget https://raw.github.com/doceme/py-spidev/master/spidev_module.c
wget https://raw.github.com/doceme/py-spidev/master/setup.py
sudo python setup.py install
```

Once the installation is done, we are good to go for writing the code for our sensor station. To understand the code, you first need to understand the data to be sent to the MCP3008 in order to get the right data. As mentioned earlier in the introduction of MCP3008, we know that in MCP3008, there are two modes available: the single-ended mode and the differential mode.

Setting up the chip select and clock is taken care of by the `spi-dev` library, but we need to understand the data transfer operations of MCP3008. How can the IC understand that it has to perform a single-ended or differential-ended function? For that, we need to send a control command from the master, which is our RasPi itself. We need to send a nibble (4-bit) of data to MCP3008 as a command that has to work on a single-ended mode and has to provide data from the x channel. Take a look at the following table and understand that most significant bit of a nibble has to set 1 to work in a single-ended operation:

Single-ended mode (D3)	D2	D1	D0	Channel to read
1	0	0	0	1
1	0	0	1	2
1	0	1	1	3
1	0	1	0	4
1	1	0	1	5
1	1	0	0	6
1	1	1	1	7
1	1	1	0	8

Before initiating the data transfer from the RasPi to MCP3008, we need to set the CS pin low, and it should be kept low throughout the conversion process. Between two conversions, the CS pin must be pulled high for optimum performance. Fortunately, the spi-dev library takes care of this, and we have to focus more on our logic than writing our own library.

There are two functions in the spi-dev library for sending the data to SPI slave devices. One is spi.xfer(), which works on active high chip select. The other function keeps the **chip select (CS)** low during operation, and makes the CS high between two blocks of transfers. We will use the spi.xfer2() function, since our device works on active low chip selects and requires the CS pin to be made high after each conversion.

The data flow should be like this:

- Start bit, (1-bit length, State: High)
- Control bit, (4-bit length: D0 D1 D2 D3)
- Wait for the data

This is how our data will look once we send the request to read channel 1 over the SPI line:

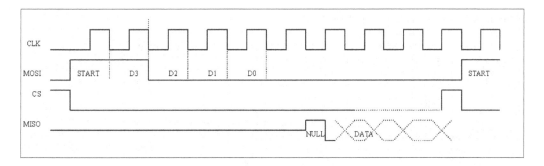

Now, the spi.xfer2([value1, value2, value3]) function sends 3 bytes over an SPI MOSI line and returns the 3 bytes back in an array. Therefore, we have to send the first byte as (00000001)b or (1)d to the SPI slave over the MOSI line. The second byte should be the nibble data of the command to the slave for selecting the channel, which is (1 D3 D2 D1)b or (8+x)d, where x is the channel number. To make this nibble a byte, we will shift the value of (8+x)d by 4 bits. As per the the timing diagram shown earlier, the third byte is not required to be sent. Therefore, we will send the third byte as (00000000)b or (0)d.

Once the command is sent by spi.xfer2(), all we have to wait for is the response from the slave, which is taken care by the spi.xfer2() function itself. It starts sensing the MISO line after sending the commands over the MOSI line. MCP3008 starts sending the 10-bit data through the MISO line and the data is stored in an array of three elements, each of length 8 bits (1 byte). We prepare a function that can be called by passing the value of channel, and get the read data back from the function as a return value. We can write that function in Python as follows:

```python
def readadc(channel):
value = spi.xfer2([1,(8+channel)<<4,0])
read = ((value[1]&3) << 8) + value[2]
return read
```

The least significant 8 bit is stored in the value[2] element of array and the remaining two bits are stored in the value[1] element. To make all other bits zero, we will perform the AND operation with (00000011)b or (3)d. Let's take an example to better understand the received data:

Received data = 00000000 01000010 01100010

The array stores the data like this: value [0] = 00000000, value [1] = 01000010, value [2] = 01100010.

We are interested in the last two bits of the value [1] element, as the ADC sends 10 bits of data. We perform the AND operation with 3 bits:

```
01000010 & 00000011 = 00000010
```

The sixth bit was already a glitch or unwanted data, which is removed by this operation. We have to shift the values left by 8 digits, as it will finally affect the decimal number.

```
00000010 << 00001000 = (10000000)b = (127)d
```

Add this shifted value to the value[2] element.

127 + 98 = 225 is the data received from the ADC convertor.

Using your sensor station – make a temperature logger

Now that you know the trick of reading the ADC value from MCP3008 through an SPI, line we are ready to interface one of the sensors with MCP3008's channel 0. You can now use your sensor station to interface with up to eight sensors in parallel. To simplify the experiment, we are going to interface the temperature sensor with the sensor station board. Imagine that the RasPi has been connected at a remote place to log temperature data. Depending on our application, once in a month or after a certain time, we need the data to be recorded manually. To log this, we will make use of Python to store data in a text file, and read these values later on for analysis.

Know the LM36 temperature sensor

LM35 and LM36 (also known as TMP36) provide the linear response of temperature changes reflected on voltage. The change in output voltage of the sensor is directly proportional to the temperature change experienced by the temperature sensor. LM35 and LM36 are centigrade temperature sensors. Because of its wide availability, these sensors can be bought online or from any retail shop near you. We will build our project with the LM36 sensor; if you have got the LM35 sensor, don't panic! The difference between these two sensors is as follows: to measure subzero temperatures using LM35, we have to provide negative voltage at the output pin of the sensor, which needs special power supply requirements. If you provide only positive voltage supply to the LM35 temperature sensor, it will sense only the positive temperature range. This is not the case with the LM36. When provided with positive voltage, it gives us the full range from -50 ˚C to +150 ˚C. It can be said that LM36 is an improved version of the LM35 temperature sensor. Let's take a look at the pin out functions and the package of the sensor. The following figure is a representation of the LM36 temperature sensor from the bottom:

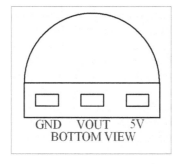

There are three pins protruding from this sensor. Consider the semicircular shape to be on top. Now the left-most pin is the GND pin, which needs the common ground. The pin in the center is the analog output pin, which should be connected to the ADC module input channel. To the rightmost pin, we can supply +2.7V to +20V. For our application, we will be sharing the +3.3V output of the RasPi expansion header (P3) to the GPCB.

Take three female-to-female jumper wires and connect them with the sensor station. Take 3.3V from header P3 and connect it to the supply pin of LM35 or TMP36. Connect the output pin to the channel 0 pin of header P2. Connect the ground wire pin to the ground pin header P4. Other sensors can be interfaced in the same manner. Longer wires can be used to put the temperature sensor in an application area to keep the sensor station away from the application.

Write the application

Now we are ready to write the code for the interfaced temperature sensor on our sensor station. We will use the same function to read the channels of MCP3008 one by one. If you have interfaced many sensors at a time, you can use the same code in your application. Along with that, we will log the data into the text file using another function, which will be introduced while explaining the code. While staying in the same folder, mcpgeneric, open nano editor using the sudo nano sensorstation. py command and start typing the following code:

```
import spidev
import time
import os
import csv
#start the SPI bus by opening the spi port
spi = spidev.SpiDev()
spi.open(0,0)
#SPI port 0 opened and Device Chip Select set to 0
#function to read the channels of MCP3008
def readadc(channel):
    value = spi.xfer2([1,(8+channel)<<4,0])
    read = ((value[1]&3) << 8) + value[2]
    return read
#writer is an object or file reference to .csv file
writer = csv.writer(file('Datalog.csv','ab+'))
```

```
while True:
    #creating the list for the different values of each channels
    datalist = []
    for i in range(0,8):
        #read channel one by one using range of 0 to 8
        data = readadc(i)
        #append data into the datalist created
        datalist.append(data)
        #convert temperature value from data received
        temperature = ((data * 330)/float(1023))-50
        print temperature
        time.sleep(3)
        print (datalist)
        #write data into the file 'Datalog.csv'
        writer.writerow(datalist)
```

If you still haven't connected the sensor station with the RasPi, connect the sensor station board using the ribbon cable. Here is how the code is structured: at first, we have to import the useful libraries and APIs to call when the program runs. The spare imported libraries are spidev, which provides SPI functionalities to transfer the data, and CSV (comma separated values), which provides functionalities to import and export the data into databases and spreadsheets. Values (0, 0) in the spi.open() function define the processor's SPI port number and chip enable value, which is set by default when the code begins to execute. After the calling of libraries, we used our generic function to get data from MCP3008 through the SPI port. Then we created the object reference to the file we are creating. The file will be created in the same folder where the program is kept and saved. The ab+ parameter is used to append the new data and is abbreviated as append in binary mode.

Inside the while loop, the datalist[] array is created to temporarily save the data into, and then the last line of the program is used to update this datalist[] array in the .csv file. By passing the values to the function in range of 0 to 7 (8 will be excluded when the loop will run) using the for loop, we will scan each channel of MCP3008. As stated earlier in this chapter in the *Math behind ADC* section, the formula of conversion for temperature sensor is created, and we have converted the data value into temperature values.

If an LM35 sensor is being used, remove -50 from the formula because LM35 cannot measure subzero temperatures unless provided with a negative voltage at the output voltage line.

Currently, it is unknown what types of sensors we are going to interface. Once the sensor is interfaced, the conversion formula should be created and imported to the same code. To append the converted data, simply use `datalist.append(variable)`, and the value will be stored in the data file separated by a comma in a single row. Every sequence of the `for` loop will generate one row. You can play around with different values and conversions by interfacing some sensors on your sensor station. Have fun; it is so simple!

 You can use `crontab` the same way as you used in *Chapter 3, Measuring Distance Using Ultrasonic Sensors*, to put this Python code in startup.

Summary

Starting from the basics of analog-to-digital convertors, you gained knowledge of a process to be followed while working with ADC modules. You understood how critical it is to handle the analog data and why processors need separate cores to process this incoming data. We got a glimpse of a MCP3008 IC and its functionality. We made our own sensor station, which can be seamlessly integrated with any RasPi model. Then, you understood how the SPI functionalities can be ported to Python code and managed to get the data into the digital form. Data logging can now be done on `.csv` files to post-process the data. With this chapter, sensor interfacing and logging has become simpler than ever. Interface as many sensors you want, try with different sensors, and deploy sensor stations everywhere to retrieve the data.

In this chapter, we collected offline data, which can be post-processed after the data is logged. In the next chapter, we will upload the data online to see real-time graphs from remote places. Also, we will enable our RasPi to send e-mails to desired e-mail IDs with, with sensor data appended. This will be your first step to creating your own Internet of Things product. Get ready for this exciting project!

6
Uploading Data Online – Spreadsheets, Mobile, and E-mails

So far, we have observed the data calculated by the RasPi, but we did not take care of the data. It has been generated, observed, and discarded after making some decisions. In the previous chapter, we logged some of the data in `.csv` files, which looked like an old school way to record the data. This data is very useful when it is logged to the Internet. How would it be if we set up the RasPi to send data every second from our home and to see the graph of the parameters from any corner of the world? How would it be if you were notified by an e-mail if there were any critical conditions? Wouldn't it be nice if the analysis of the data were sent to you by an e-mail at the end of the day?

I know that we are always worried about our home when we are on holiday or even in the office. There could be a lot of data to take care of when you are away from your place, for example, temperature, humidity, motion, fire alerts, unauthorized intrusion, energy consumption, device states, and so on. There are infinite possibilities to turn your home into a smart home using the amazing RasPi.

In this chapter, you are going to learn how all these things can happen at one go. You will understand that why the **Internet of Things (IoT)** is the discussion around the breakfast table for every technology geek. You will also learn how communication usually takes place and the way the sensor nodes and devices attached to the IoT work. You will get to know why data analytics, cloud storage, and cloud access technologies are in boom. On one hand, we get the facilities, but on the other hand, there are some security concerns that should be addressed while deploying this technology. You will get some knowledge on how severe these security concerns are.

We developed a sensor station in the previous chapter. We will use this sensor station to collect the data from different sensors while doing the hardware setup. In the next step, we will upload the data to the Google spreadsheet by pushing it online, and see the real time graph of the sensor data on desktops and even on Android and iOS mobile phones. Then we will set up the RasPi to send us an e-mail of the data collected in an entire day, and also the minimum and maximum values reached by the sensor. It will also be able to send an e-mail when there is a critical situation at home.

This chapter sounds so much fun—learning by giving intelligence to the RasPi to observe the sensor data online. Let's start the journey by understanding the IoT.

Internet of Things

IoT has started attracting all scientists, innovators, technologists, engineers, and investors to start development and take part in one of the fastest growing markets in the world. It can fulfill the vision of all of these people to make the world better to live in. Once implemented on a large basis, it will make our lives easier, more comfortable, and safer. This revolution began by redesigning and replacing the smallest and basic things we use in our day-to-day lives. It is said that there will be about 30 billion devices as a part of IoT by 2020, excluding mobile phones and tablets. From toothbrushes to shoes, wallets, washing machines, switches, vehicles, and much more, they all will be part of IoT.

Some of the IoT devices learn our everyday habits, adopt them, and then respond by taking the decision by themselves. It can be really amazing to see the revolution in every physical entity by adding communication capabilities to it. There are no particular definitions of IoT. Technically, we can understand IoT as adding Internet connectivity (basically a unique Internet address), and also the capability to sense everyday physical objects or entities that can communicate with each other and with mobile devices, such as phones and wearables. Some people in communities believe that the IoT space is overloaded as it has a lot of expectations and adds a lot of redundant devices, which are really not required to add comfort to our lives but are forcibly added.

 One of the interesting projects on artificial intelligence and IoT is Jibo, a family robot. Take a look at http://www.jibo.com.

IoT can be represented as an integrated fabric of devices, networks, data, calculations, analytics, and people as essential elements. We can imagine an IoT network as a big web mapped in a home to connect all devices, such as lights, fans, toasters, thermostats, geysers, switches, refrigerators, ACs, and media devices. Imagine a situation where you arrive at the airport after a long summer holiday and the temperature at home is intolerable due to no presence at home. Once your mobile phone connects to the Internet, the AC in your home will get to know that you have arrived back in the city. It will estimate the time of your arrival according to the traffic in the way and set the rate of cooling your home. In addition to that, it will know what temperature to set according to your daily habits. In the reversed situation, when you are leaving for your holiday, the device in your home knows the location of your mobile phone, and it will disable all the unnecessary devices as set in the mobile phone's application. This can also be controlled manually through mobile applications. IoT technology is applicable not only in homes but also in hotels, hospitals, industries, offices, and transportation—in short, everywhere. Really, it will be the Internet of everything.

Sensor nodes

As we know, there are many parameters to sense at a single place. We have already built a sensor station, which can ultimately be called a sensor node once installed in a corner of our home or office. Then it will be silently observing the surroundings and storing all the data. Imagine connecting that node to the Internet, and many other nodes as such. A sensor node can be of any size, for example, starting from the size of a table top projector down to the size of a dust particle! So, can we call it smart dust? Yes! Researchers have already implemented smart dust particles (also called motes) that get energy from light, temperature, or a thin film battery. Widely used sensor nodes range in size from a fingertip to a palm. These sensors can create a wide network over a targeted area, and can number from hundreds to thousands.

Implementing such sensors may face several challenges. These challenges start from the hardware end. Most of the devices in the network will be using low-cost MCUs to process the sensor data locally and transmit it to nearby devices or to an Internet gateway. These low-cost MCUs will require enough RAM to run security protocols and the user application. In addition, the energy consumption of such MCUs must be as low as possible to run the node on a 3V button cell for 4-5 years. This can be achieved by putting the processor in sleep modes between periodic transmissions of data. The duty cycle of such processors can be near 1 percent or at the required periodic rate, according to the calculations of data. Putting the processor in sleep mode and resuming should take no more time than a fraction of the set duty cycle. There are PLLs and clocks, which take time to become stable, and the stability of a processor core is solely dependent on these clocks. This stability time (can be called settlement delay) can go up to milliseconds, which can ruin the dreams of the hardware designer to run the device for a longer time on a single cell. There are many other hardware issues to be addressed by a hardware developer.

Due to many hardware constraints, there are many challenges faced in the software end too—utilizing the proper RAM space, clearing it periodically, and writing efficient algorithms with the communication protocol stacks in such a limited amount of RAM. The next thing can be the remote **firmware upgrade over the air (FUOTA)**. It may be the case that the node needs an immediate firmware fix after deployment in the field to fix the cracks in the received data. A firmware upgrade can also fix the performance of the hardware by readjusting the handling of interrupts and sleep timings, and optimizing the code size and efficient usage of RAM. FUOTA ensures that the deployed sensor node works properly and responds well to other nodes. These firmware upgrades are maintained under the version control. Special handling is required on the exceptions, and all possible failures are considered while making the device remotely upgradable over the air. For example, suppose the node is being upgraded, and the power level in the device is so critical that it leads to shutdown of the device. When it resumes on arrival of power, it should revert to the previous stable version of the firmware and ask for the latest firmware upgrade. Download time, download size, and timings of the day are other factors that affect the success rate of FUOTA.

Communication

Communication between sensor nodes should take place through a lossless medium. Packet losses play a major role in failure of data transmission. Communication is the most essential part of the IoT infrastructure. It is a duplex medium to collect and send data as well as commands. After a connection has been established, it has to be maintained and managed all the time.

Communication of all of these devices follows a protocol to establish a link between each other. A protocol defines a set of rules to be followed by such nodes so that all the critical conditions that may disrupt the communication can be avoided. A protocol also provides a well-defined format for sending the commands over the network. In previous chapters, we used wired protocols such as SPI. Along with the rules that the wired protocol has, wireless communication needs set of rules to be applied on parameters such as frequency, addresses, timings, sequences, frame types, and many more parameters structured in a way to utilize the resources.

Implementation of IoT will require millions of unique identification addresses. This would require IPv6 to be implemented on these devices. Providing an address for a device and making an interactive network are not the only challenges. The real challenge is to handle the data in an efficient and secure way. Several communication protocols used in IoT space these days can be enumerated as Wi-Fi (2.45 GHz, 802.11 a/b/g/n), Zigbee (2.45 GHz, 802.15.4), Bluetooth 4.1 LE (2.45 GHz bluetooth low energy, iBeacon), Z-Wave (900 MHz, sub-1 GHz RF), 6LOWPAN (IPv6 over Low-power Wireless Personal Area Networks), and many more. There is rapid progress being made in developing and improving these protocols to be more energy efficient and robust. There are daughter boards available for the RasPi for adding these functionalities by interfacing these add-on boards. There are many start-ups that are now involved in developing products based on these protocols.

The cloud

One of the greatest advancements in technology experienced over the past few years is the cloud computing technology. Embedded systems and IoT, being low in cost and small in size, can take maximum advantage of this technology. These embedded devices have memory in the order of a few KBs. Sensor data collected by such a system can be stored, but this data can cause an overflow in the memory after a particular amount of time. It can be very difficult to store long sensor data. Other constraints with these devices can be low processing power while being in low-power operation to run batteries for years. We can get rid of these constraints by deploying cloud services on our sensor networks. The entire data, uploaded by several sensors, can be calculated and interpreted on the cloud instead of local computation to save power and increase the sleep time of the processor. In the sensor network, a gateway can be set up, which can always be connected to the cloud to act as a gateway transceiver. All computation-intensive algorithms can be deployed on the cloud rather than putting a hefty processor to act as a hub and collect as well as send data from the cloud.

Data analytics

IoT is expected to produce a huge amount of data from diverse places that is aggregated and high-speed, thereby increasing the need to better sort, store, clean, transform, and process such data. There will be a mine of data, and we need to gather much more information from this pool of data. Disaggregation of the data and extracting meanings from it are a part of data analytics. Using data analytics, one can make their algorithms better to improve the user experience of a product. For example, a home automation system collects a large amount of data. By collecting the data, it knows when you turn on your TV or AC, what temperatures you set, and when you come home. By collecting all of this information, mining in this data, and applying machine-learning algorithms, a system can learn your habits, which can further be useful for saving energy. The best example of a connected and learning device is the Nest thermostat. Data analytics and mining will be the key areas of development in IoT. Not only functionalities but also user experience is very important in automation systems. Data analytics is nowadays used to disprove the probability theories and models derived from user habits.

Security concerns

Security is the major concern when IoT is deployed everywhere. Engineers building IoT devices leave some loose ends at the security, and that only starts creating problems. It's a noisy, nosy world! Neighbors are always interested to know your daily habits and activity. By enabling the communication functionality in all devices in your home, you are indirectly attracting a hacker to access, control, and analyze your devices. Motion sensors deployed at the door, the water heater in bathrooms, thermostats, air conditioners, and water motors could be within direct access of a hacker, by which he can know that the back door is left open while you are in the drawing room watching your favorite TV show. He can directly access and control all devices by sitting near your home or from the Internet. Now imagine a hacker intrusion in a fully automated, IoT-enabled operation theatre or in the vehicles you drive! There are multiple security protocols that are being developed and researched, with smaller footprints on RAM and ROM for low-power embedded devices. They are secure enough to protect the devices from such attacks. There is a need to develop algorithms to bridge the gap between the Internet and the IoT space.

Hardware setup

With a lot of information on IoT provided in the previous section, we are going to make a small IoT project of our own this time. This hardware setup will be a very easy setup compared to the previous chapters. The hard work we did in *Chapter 5*, *Using an ADC to Interface any Analog Sensor with the Raspberry Pi*, will be very helpful to us. We will be using the sensor station to get the data of the temperature sensor. Listing down the needs for this chapter, we will require the following:

- A sensor station
- A temperature sensor (LM35/TMP36)
- A GPIO ribbon cable for Raspberry Pi model B (26 pin), A+ (40 Pin) or B+ (40 pin), or RasPi 2 model B (40 Pin); female-to-female
- Male-to-female and female-to-female jumper wires
- A Raspberry Pi, a power adapter, an Ethernet cable, a personal computer

Connect the sensor station using the ribbon cable. Power up the RasPi with the power adapter, and start the session in PuTTY using your personal computer. Cross-verify that your computer is sharing the Internet connection with the RasPi. Use the following command to check:

```
ping -c 4 www.google.com
```

If it is not sharing the Internet connection, go through the guide provided in *Chapter 1*, *Meeting Your Buddy – the Raspberry Pi*. Obviously, connecting the RasPi to the Internet is very important. Remember the Internet of Things! Let's dig into the software side. There's much more to look in the software now.

Synchronizing the clock with the Internet

It is really difficult when the date and time of the RasPi are not synced with the local time zone while logging the data. It is much more difficult once you set up your logging device or sensor node in a remote place that doesn't have uninterrupted power supply. You might have noticed that whenever the RasPi boots up, the clock shows the incorrect time. Check it yourself by entering the `date` command in the command-line interface of the RasPi. Also, in the previous chapter, you might have noticed that the time logged in the `.csv` log file is not showing the refreshed date. No, the RasPi doesn't time-travel when it's sleeping! It's just because it doesn't have a dedicated button cell to power its **real-time clock** (RTC). To make the RasPi cheaper, the designers removed a lot of functionalities that normal desktops or laptops have.

A desktop computer has an inbuilt button cell to power the internal RTC. To achieve this with the RasPi, you can interface an IC called DS1307, which communicates over the I2C protocol. Then you can fetch the time from the RTC whenever you want. You can power this IC with the CR2032 button cell. Illustrating the interface of this hardware may require another section, and it is beyond the scope of this book. We can use a simpler way of synchronizing the clock since we have an Internet connection on the RasPi. We will use the **Network Time Protocol (NTP)** service to provide the date and time for the RasPi. For a long time, this protocol has been used to keep time synchronization between two computers or servers. On Wheezy Raspbian, the NTP client is already installed. All you need to set is the daemon at startup, which runs in the background, to fetch the timings from the Internet servers. Type the following code to set up the daemon:

```
sudo nano /etc/rc.conf
```

You may see the blank configuration file, in which you can write this:

```
DAEMONS=(!hwclock ntpd ntpdate)
```

In the preceding command, the ! mark means that the hardware clock will remain as it is, but we will update the software clock in the background. Press *Ctrl + X*, then *Y*, and then the *Enter* key to save and exit the configuration file. Then reboot the RasPi by entering the `sudo reboot` command. This daemon will be silently running in the background.

Upon reboot, just check the system time by entering the `date` command, and tally it with the current time and date.

If you did not set the time zone of the RasPi, then you might see the wrong date and time displayed by the `Date` command. To set the time zone, you can enter one of the following commands:

Enter `sudo dpkg-reconfigure tzdata` and follow the user interface with the keyboard.

Alternatively, you can set the same configuration you did at the first setup by commanding `sudo raspi-config` to open a RasPi configuration menu, and then navigate into the international locale settings. Reboot the RasPi once you configure this.

Uploading data on Google spreadsheets

We are all set with the synced time to the local time zone. We can create an application that does the same data logging as we did in the previous chapter, but this time it is different and cool. We will upload the data on Google spreadsheets, and the data can then be accessed from any corner of the world. You can monitor home temperature and humidity by sitting in your office. Either you can grab the project you did in *Chapter 4, Monitoring the Atmosphere Using Sensors*, and observe the temperature and humidity with the DHT11 sensor, or you can just bring the sensor station built in the previous chapter using MCP3008.

Before we get into the code, we need to prepare the things listed here. Use your personal computer's Internet browser to perform these steps:

1. Log in to your Gmail account or create a new one.

2. Open or link Google Drive by logging onto `http://www.drive.google.com` with your Gmail ID.

3. In the user interface of Google Drive, on the left-side panel, click on the **New** button and select **Spreadsheet** from the drop-down menu. You will be redirected to a new tab.

4. In the new tab, name the spreadsheet `Logging`. Avoid blank spaces in the name to reduce errors, as we are going to use this name in our code.

5. Click on the **Add-ons** drop-down menu from one of the spreadsheet menus, and then check out **Get Add-ons...**. In the search bar, type `Remove Blank Rows`. Click on the free button and install it.

6. In the first row of the spreadsheet, enter the column titles as `Time` and `Temperature`.

7. Click on the **Add-ons** menu again and then on the **Remove Blank Rows** add-on. Check out the **Delete or hide blank rows** option. You will be able to see the sidebar with some options. Click on the top-left corner of the spreadsheet to select all rows and columns. Then, in the sidebar, select the **All row cells must be blank** option and delete all blank rows. It should look like what is shown in the following screenshot.

 The reason for doing this is that the code will start entering the data from the last blank row of the sheet by creating a new row. If you have a thousand rows, it could be difficult to manage the data as it will start entering data from 1001st row.

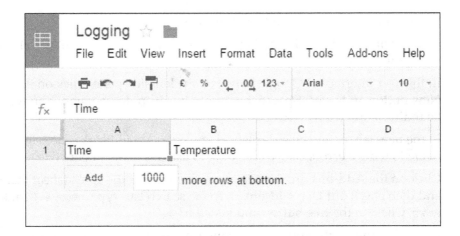

8. Close the tab, and we are ready to code in the RasPi.

Now, on the RasPi, install the library that provides support to push the data to the Google spreadsheets. Open a PuTTY session to enter these commands to install the `gspread` library:

```
git clone https://github.com/burnash/gspread.git
cd gspread
python setup.py install
```

It is really important to install the `gspread` library, as it gives us the handy functions to upload the data on the desired spreadsheet document on the selected sheets, cells, and much more.

Here, we are going to use `MCP3008`, as it can be useful to interface as many sensors as we can. We will recall the functions used to get data from the SPI port:

```
#start the SPI bus by opening the spi port
spi = spidev.SpiDev()
spi.open(0,0)


#function to read the channels of MCP3008
def readadc(channel):
    value = spi.xfer2([1,(8+channel)<<4,0])
    read = ((value[1]&3) << 8) + value[2]
    return read
```

The explanation of fetching data using the `readadc()` function has already been given in the previous chapter. We will use the same generic function to fetch the data from the sensor and send it over the Internet.

Observe the following Python code and get an understanding from the description given just beneath it:

```
import os
import spidev
import glob
import time
import sys
import datetime
import gopread

#start the SPI bus by opening the spi port
spi = spidev.SpiDev()
spi.open(0,0)

# Enter your account details (Your Gmail ID and Password) as shown
here
email = 'gajjar.rushi@gmail.com'
password = 'raspberrypi'

#Name of Spreadsheet created in Google Drive
spreadsheet = 'Logging'

#Putting the exception call in python to attempt for logging in Gmail
```

```
try:
    ret = gspread.login(email,password)
except:
    print('Oops! Check Internet Connection or Login Credentials')
    sys.exit()

#open the spreadsheet by either of these two options
worksheet = ret.open(spreadsheet).sheet1
#or with the spreadsheet key
#worksheet = ret.open_by_key('1eQth-TY4FXFKChB5RFPhelQ6zn47NWDESh13Wk
XGQAk')
#prefer First Option

def readadc(channel):
    value = spi.xfer2([1,(8+channel)<<4,0])
    read = ((value[1]&3) << 8) + value[2]
    return read

while True:
    #Get data from Channel 0, TMP36 Temperature Sensor
    val = readadc(0)
    temperature = ((val * 330)/float(1023))-50
    values = [datetime.datetime.now(), temperature]
    worksheet.append_row(values)
    time.sleep(5)
```

At the top of the code, you can see that some libraries are imported for basic OS functionalities and to fetch the date and time. After importing these libraries, the SPI port is opened to fetch the data from MCP3008. Next to the SPI port, we will assign variables to store the Gmail ID and password. These values must be the same as your Gmail login credentials. Then the Python code tries to log in to your Gmail account using the e-mail ID and password variables. Exceptions are best handled by the try-except functions. The code should not hang somewhere if there is no Internet access or login access to the RasPi. Rather, it should show that something is wrong. You will understand the try-except function in detail when we discuss e-mail notification in upcoming sections.

After successful login, it searches for the spreadsheet named **Logging** and selects **sheet1** in the **Logging** spreadsheet. If you will observe the link to the spreadsheet in the web browser in your desktop computer, you will see a long, random number stated as 1eQth-TY4FXFKChB5RFPhelQ6zn47NWDESh13WkXGQAk. This is the key to your spreadsheet. However, you should preferably select the sheet by the ret. open() function as it is a direct and clean way to access the Google spreadsheet.

On calling the function of the SPI, the temperature value will be stored in a variable, which will be then passed to the worksheet.append_row() function to be pushed to the Google spreadsheet.

By running the code, you will be able to see that a new blank row is automatically created and then the temperature value is added, with the latest date and time. Add a graph in the `Temperature` column and see the live feed updated in the `Temperature` column.

 To add a graph, click on **Insert**, select the **Chart** option, and type `Sheet1!B:B` in the data range textbox. Give titles to the horizontal and vertical axes and select the type of chart.

I took some ice cubes and a matchbox to simulate the results, and the data feed looks amazing, as shown in the following screenshot:

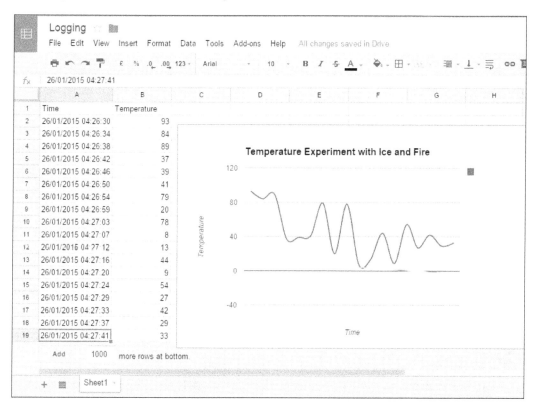

This experiment will give you a kickstart to interface more sensors. Interface two or more sensors now and you can append the data in the third and fourth columns using the following function:

```
values = [datetime.datetime.now(), value1, value2, value3]
worksheet.append_row(values)
```

That's all! You can share this sheet (by clicking on the **Share** button located at the right corner in the Google spreadsheet and adding the Gmail IDs) with your friends and cousins living far away from you so that they can just check out this cool new featured product developed by you.

Live feed on mobile phones

You can install the Google sheets application in your Android- or iOS-based phone and just see the feed coming from the RasPi on your mobile phone. Just download it from the Play Store or an iStore (the App Store) and enjoy the live feed right from your office. My Android phone's screenshot is as follows:

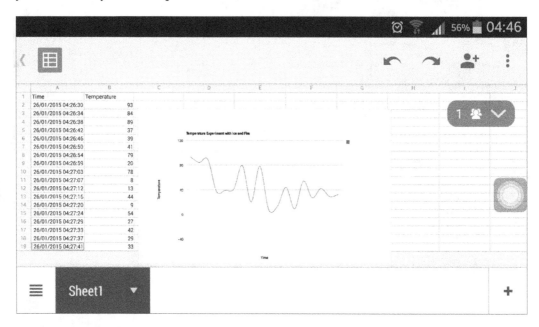

Getting notified by e-mails

Even after accomplishing the previous project, it still feels as if something is missing. Why should we continuously look at the data and check whether there is a critical condition or not? We will therefore add the following functionality: every night at 12:00 a.m., the RasPi will send an e-mail stating the minimum and maximum temperatures of the day. We will get the warning e-mail by setting a threshold value of temperature whenever the critical condition arises. Suppose 30 degrees Celsius is the limit, and increasing the temperature beyond that may be the sign of a critical condition at your home.

Here, we are going to use an inbuilt Python library called `smtplib` to get the functionalities of sending e-mails from the RasPi. The `smtplib` module outlines the **Simple Mail Transfer Protocol (SMTP)** client, which can be imported to send an e-mail from a client to any Internet machine using the SMTP daemon. The `mimetypes` module provides us with the functionality to convert a URL or file type to **Multipurpose Internet Mail Extensions (MIME)**. It is a functionality that lets people use the protocol to exchange data over the Internet. Now import e-mail and the `mime.application` modules, which can give us the e-mail functionality, such as sender e-mail ID, receiver e-mail ID, subject of the e-mail, attaching a body to an e-mail, and much more. The `sys` library provides system-related functions and parameters, such as exiting the attempt of the `try` exception and continuing the code. The difference between the `datetime` and `time` libraries is that `datetime` provides the software clock with date and time data strings, while `time` provides us with the `sleep` function to halt the code for some time.

The outline of the algorithm is pretty simple. It is useful for getting an understanding of the code written in Python:

1. Import the essential libraries.
2. Open the SPI port.
3. Define the SPI data fetching function.
4. Define the e-mail functions and pass all the important variables (such as `sender_email`, `sender_password`, `receiver_email`, `subject`, `min`, and `max`) by calling this function with the right parameters. The e-mail will be sent.
5. Start a forever-running loop.
6. Create a `.csv` file with a name appended with today's date and time.
7. Store the current date in a variable called `prev_date`.
8. Set the temperature threshold limit.
9. Under the forever-running `while` loop created in the fifth step, put an indent and create one more `while` loop to check the current date change.
10. Call the sensor-reading function to get the value from `MCP3008`.
11. Get the minimum and maximum values of temperature from the `.csv` file.
12. Define two sets of conditions in which the RasPi will send an e-mail, as follows:
    - When the current temperature goes beyond the `Temperature` limit
    - When the date is changed

13. Keep updating the `.csv` file locally and change the file when the date changes.

14. Start again from step 10.

This algorithm gives us an overview of how the code will flow and the elements to be inserted. Check out the following code:

```python
import smtplib
import mimetypes
import spidev
import email
import email.mime.application
import sys
import csv
import datetime,time


#start the SPI bus by opening the spi port
spi = spidev.SpiDev()
spi.open(0,0)


#Defining the ADC Reading function on SPI port
def readadc(channel):
    value = spi.xfer2([1,(8+channel)<<4,0])
    read = ((value[1]&3) << 8) + value[2]
    return read


#Email Function to send the email by passing the correct parameters
def email_data(sender_email,sender_password,
receiver_email,subject,min,max):
    mail = email.mime.Multipart.MIMEMultipart()
    mail['Subject'] = subject
    mail['From'] = sender_email
    mail['To'] = receiver_email
    content = email.mime.Text.MIMEText(""" Min temperarute
temperature is %d, Max temperature temperarute is %d"""%(min,max))
    mail.attach(content)
    s = smtplib.SMTP('smtp.gmail.com:587')
    s.starttls()
    s.login(sender_email,sender_password)
    s.sendmail(receiver  ,[receiver], mail.as_string())
```

```
        s.quit()

while(1):
        #create a new csv file
        #create a new csv file
        filename = 'Sensor_data_' + str(datetime.datetime.now()).split('
') [0]
        writer = csv.writer(file(filename+'.csv','wb'))
        prev_date = str(datetime.datetime.now()).split('.')[0].split(' ')
[0]
        max_sensor_data = -9999
        min_sensor_data = 9999
        #setting the temperature threshold limit to 30 C
        temp_limit = 30

        while(1):
                cur_datetime =str(datetime.datetime.now()).split('.')[0]
                cur_date = cur_datetime.split(' ')[0]
                sensor_data = readadc(0)
                min_sensor_data = min(sensor_data,min_sensor_data)
                max_sensor_data = max(sensor_data,max_sensor_data)

                #check if the threshold value is crossed
                if( sensor_data > temp_limit):
                        try:
                                email_data('gajjar.rushi@gmail.com',
'raspberrypi','gajjar.rushi@gmail.com','Warning !!!',
min_sensor_data,sensor_data)
                        except:
                                pass
                #check if the date is changed, if changed: send email
and start from creating new csv file
                if(cur_date != prev_date):
                        try:
                                email_data('gajjar.rushi@gmail.com',
'raspberrypi','gajjar.rushi@gmail.com',filename,
min_sensor_data,max_sensor_data)
                        except:
                                pass
                        break

                #update the csv file
                writer = csv.writer(file(filename+'.csv','ab'))
                writer.writerow([cur_datetime,sensor_data])
                time.sleep(1)
```

Let's understand the e-mail function first. In the e-mail function, we are passing variables (`sender_email`, `sender_password`, `receiver_email`, `subject`, `min`, and `max`) that come when an `if()` loop's condition is satisfied. In the `content` variable, we are storing the body of an e-mail containing the minimum and maximum temperature values. Then it tries to log in using the e-mail ID and password provided in the functions.

The code flow starts from the first `while` loop, which will start running forever. This outer `while` loop will complete one cycle every day because we are putting an inner `while` loop to run forever and break when the date changes. If you look closely at the `filename` variable, you will see that it sets the filename as `Sensor_Data_` appended with the current date. The `datetime.datetime.now()` function provides the time and date with precision in milliseconds. Millisecond precision is not required for our project and application. The `Splitsplit()` function removes the milliseconds from the raw data. Once we get the filename, we save it as a `.csv` file using the `csv.writer()` function.

A few examples of the `split` functions run in Python Shell returned these values:

```
>>> str(datetime.datetime.now())
2015-01-26 17:12:49.998000
>>>str(datetime.datetime.now()).split('.')[0]
2015-01-26 17:13:17
>>> str(datetime.datetime.now()).split('.')[0].split(' ')[0]
2015-01-26
```

To detect the date change, we will use the `split()` function two times to extract the date from string returned by the `datetime.datetime.now()` function, and it will be stored in the `prev_date` variable.

We are using *max_sensor_data = -9999* and *min_sensor_data = 9999* to prevent false data segregation in the `min()` and `max()` functions, which are used to find minimum and maximum values from the `.csv` file. The `temp_limit` variable can have your desired value—the value at which you want to send an e-mail from the RasPi.

It would be good to add `try`, which handles the selected exceptions. As explained in the previous sections while experimenting with Google spreadsheets, it works on a simple principle:

1. Execute the `try` clause. If no exception occurs, the `except` clause is skipped and execution of the `try` statement completes.

2. If an exception occurs during execution of the `try` clause, the next part of the clause is skipped. Then it goes to the `except` clause and it gets executed.

The rest of the code is self-explanatory and can be understood from the comments provided in the code itself. You can set the same sender and receiver e-mail while you are testing your code. Every day, data is stored in a `.csv` file with `date` as a name, so whenever you open your RasPi you can just take out an entire folder of information packed with the `.csv` files.

Once you code, you can properly fit the sensor station in an airtight container along with the RasPi, and fix it near to the router of your home. Connect an Ethernet cable directly from your router to get an Internet connection for your RasPi. Your computer being in the same network, you can ping the IP address of RasPi to check whether it has successfully connected to the router and your network.

Integrating everything

Now it's your turn to integrate the Google spreadsheet project with the e-mail notifier project. All you have to do is better arrange the events of sending data to Google spreadsheets, logging data locally, and sending the e-mail. You can even try to attach the `.csv` files when you send a day's summary. The application of this project is so vast that you can do much more with this code by adding functionalities. By adding multiple sensors with multiple RasPi boards in different corners of your home, you can really get much more data and many more alerts from your home.

If you are installing a RasPi in a remote place, do not forget to add the Python script in `crontab` as described in *Chapter 3, Measuring Distance Using Ultrasonic Sensors*, to start logging the data again when the RasPi boots back.

Common problems faced

While building a project, there may not be many problems faced as it is mostly a development in software. Still, mistakes can happen in the ways described here.

In your code, do not declare a variable name or function name same as Python data type or Python prebuild function; for example, naming a function `email()`. This is not valid and can create a runtime error in the code. The same problem can occur when you set the Python script filename as a predefined Python function, for example, `email.py`.

While experimenting with gspread, try to write the script in the install library folder of gspread itself. Alternatively, you can play with Linux to make the function and the file available as an external location of the files. For example, we installed gspread using these commands:

```
git clone https://github.com/burnash/gspread.git

cd gspread

python setup.py install
```

In the same gspread folder, write the Google spreadsheet data logging program. One of the most common issues with Python is the indentation of the code. Most errors happen while writing the loops. To understand the e-mail notification code better, just take a look at the snapshot of the code in the following screenshot. Ignore the . sign in the text. Closely look at the long hyphens before the lines. These hyphens show the number of tabs required to write the line.

```
while(1):
    filename = 'Sensor_data_' + str(datetime.datetime.now()).split(' ')[0]
    writer = csv.writer(file(filename+'.csv','wb'))
    prev_date = str(datetime.datetime.now()).split('.')[0].split(' ')[0]
    max_sensor_data = -9999
    min_sensor_data = 9999
    temp_limit = 30
    while(1):
        cur_datetime = str(datetime.datetime.now()).split('.')[0]
        cur_date = cur_datetime.split(' ')[0]
        sensor_data = fun()
        min_sensor_data = min(sensor_data,min_sensor_data)
        max_sensor_data = max(sensor_data,max_sensor_data)

        if( sensor_data > temp_limit):
            try:
                email_data('gajjar.rushi@gmail.com','raspberrypi','gajjar.rushi@gmail.com','Warning !!!',min_sensor_data,sensor_data)
            except:
                pass

        if(cur_date != prev_date):
            try:
                email_data('gajjar.rushi@gmail.com','raspberrypi','gajjar.rushi@gmail.com',filename,min_sensor_data,max_sensor_data)
            except:
                pass
            break
        writer = csv.writer(file(filename+'.csv','ab'))
        writer.writerow([cur_datetime,sensor_data])
        time.sleep(1)
```

Other general errors include the Internet connection sharing your RasPi or login access to your Gmail. Sometimes, it may happen that you have added an extra layer of security to log in to your Gmail ID, which can prevent a second party from logging in to your e-mail. It is recommended to remove that security layer or make a cool and separate e-mail ID for your RasPi.

Summary

We started this chapter with the essentials of IoT. The information provided in the sections did not encompass the entire concept of IoT. IoT is big space, including sensor nodes, sensor networks, communication, cloud computing, data analytics, and security. There can be many more facets of IoT that require a lot more attention from developers. You came to know that IoT could be the future of Internet and data, starting from the smart dust all the way up to smart cities. Security is a major concern, which needs immediate attention and research to meet the requirements of the embedded world.

This chapter gave you knowledge on observing the data inserted into Google spreadsheets from the sensor interfaced with your RasPi. You learned how a RasPi can log in to a Gmail server on behalf of you and send an e-mail when there is a real need. We prepared a project that purposefully sent an e-mail at the end of the day, sending the analysis of a data. Minimum and maximum values of the data can be a very small example of data analysis. It was a good challenge to integrate gspread and e-mail notification codes to make the code even more practical.

In the next chapter, we are going to interface an image sensor (camera) with the RasPi and test its operability. A complete illustration on installing the OpenCV library will be provided to get you started with image processing on the RasPi. You will make a project with live video streaming to use the RasPi as a CCTV camera and capture an instance of motion in your house. It will be an amazing experience and a great kickstart for you to develop a project using an awesome library known as OpenCV.

7
Creating an Image Sensor Using a Camera and OpenCV

There has been an enormous amount of research behind a small name—image sensor. From developing CMOS sensors, which can take crystal-clear images, to employing enhancement of images using image processing, there are so many efforts involved. Making different types of cameras suitable for different applications and processing algorithms is not easy at all.

Image sensors are used in cameras, and they are very difficult to use in unpackaged conditions, as they are fragile and sensitive to electrostatic discharge. We will be using a camera (basically a lens with the protective case and an interface cable) to demonstrate image capturing using a RasPi. Performance of a camera with a RasPi is always rated low in discussions because of its poor performance with image processing and video processing, with comments that image processing and video capturing is not the RasPi's strong point. Somehow, these comments are true when we run computation-intensive algorithms with a huge amount of processing, and the RasPi gives a delayed response. But it all really depends on the hardware hookups at the time of camera interfacing. There are many factors that could affect the performance of the RasPi when you are using the camera. This was the issue with the older models of the RasPi, such as Raspberry Pi 1 model B, B+, and A+. But now, with Raspberry Pi 2 model B, this will not be the case. RasPi 2 model B is six times more powerful in performance than the older RasPi models.

In this chapter, we will start off by giving a brief introduction to image processing and OpenCV. Then we will go through the camera interfacing ports and look at the different kinds of cameras that can be interfaced with the RasPi. Next, we will do a long and tiring (but well-worth the time) installation of the OpenCV library on our RasPi. Once the libraries are installed, we will write a C program to capture an image from a camera. Moving one step ahead, we will experiment with the live streaming of a video over the same network as the one in your house. In a further step, a project will be prepared to detect human motion or movement in a particular area, capture the image, and alert you immediately.

Image processing

Have you ever tried to look at an image by zooming it to a maximum level? It just looks like a floor with organized tiles and colorful patterns on it. These square tiles in the image are known as pixels. Basically, an image is a group of such pixels, with each pixel containing a particular value of color, which forms the recognizable patterns by providing information to the human eye. It all depends on how humans perceive the image by observing shapes and colors. Each pixel in an image contains information that can be generated from a byte (8 bits) or a couple of bytes, which defines the depth of an image. Depth of an image is nothing but the number of bits present in a single pixel. Current display monitors and graphics engines support up to 64-bit depth of images. Basic types of images are binary, grayscale, and RGB, and many more such as HSV, HLS, and YCC are known types. A grayscale image does have the range of values from 0 to 255 in the 8-bit mode, while a binary image has them in the range of 0 to 1. As the name itself suggests, a binary image has only black and white colors possible on the image plane. The difference between a grayscale and a binary image can be observed in the following images:

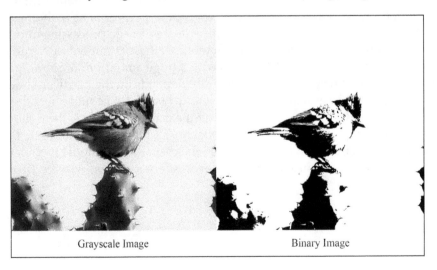

Grayscale Image   Binary Image

An RGB image has a total of three planes to access, as it has R (red) values varying from 0 to 255; and the same applies for G (green) and B (blue) colors for 8-bit depth. A grayscale image has shades from white to black, with a total of 256 discrete values (0 to 255) for an 8-bit deep image.

Image processing is performed to enhance an image to a certain levels, to extract maximum information. Just as we perform computations in signal processing, we do them in almost the same way in image processing. Here, the plane is a two-dimensional array of information in the form of pixels, whereas in signal processing, it is a one-dimensional array of signal streams. In complex image processing applications, there can be more than 20 or 30 algorithms and functions to be applied to an image to retrieve the distorted image or enhance the quality of it. In image processing algorithms, images are used as two-dimensional arrays. The number of X and Y columns of the image represents the image resolution. When algorithms are applied to an image, they individually process every pixel of it. Changes in the pixels occur according to the algorithm applied. Image processing can also be performed on a video, which is nothing but a group of images shown at a particular rate to create an illusion to the human eye, to perceive it as motion.

In a motion picture, there are multiple images (frames) showed at a time. There should be at least 22 to 25 **frames per second** (FPS) to create an illusion of motion for the eyes. These frame rates have been in practice since long. Mostly, cartoons and silent films in the 1920s started using a picture frame rolled in front of a concentric light to project it on a screen. Visual perception of every human is different. It is observed that less than 16 frames per second doesn't give better visual experience, as there will be a brief disruption of darkness. These days, the latest technologies, such as full HD (1080p), use frame rates of 50 to 72 FPS. Some applications require higher frame rates—up to 20,000 FPS—compared to our digital camera (30 FPS) or 3 FPS CCTV cameras. High-speed cameras for ultra-slow-motion videography require 20,000 FPS. All of these applications require a huge amount of computational power. Image processing on live video is mostly performed using individual frames.

Digital image processing is a vast field. Learning image processing can help you get employed in space agencies such as ISRO, NASA, or JAXA. Image processing is extensively used by space agencies to filter distorted images received from space telescopes and satellites. Image processing is used not only in space agencies but also in medical treatments, robotics, agriculture, automation, and stitching of Google maps images. Google sphere is also one of the best examples of image processing.

Understanding image processing requires a lot more studies of spatial domains and filtering algorithms. Currently, several algorithms have been developed to enhance the image quality. Fields such as artificial neural networks and pattern recognition are trending topics in research.

OpenCV

One of the most powerful and widely used cross-platform libraries for image processing is OpenCV, which can run on any hardware or software platform. It can be installed on the RasPi to get functionality such as face detection, finding contours, gesture recognition, motion tracking, 3D depth perception using two cameras, and many more complex algorithms for machine learning. **OpenCV** stands for **Open Source Computer Vision**. It was developed by Intel. It is basically written in the C++ language and can have full interfacing with Python, Java, or MATLAB. Check out `http://opensource.org` for the list of available open source libraries. With aiming for computational efficiency in mind, OpenCV is designed to work with multicore architectures. Writing OpenCV in C and C++ is common across the globe among image processing experts. OpenCV can also be written in the Python language, as the library has full development and supports functions in the Python script.

Computer vision is a quickly growing field because of cheaper and advanced cameras and inexpensive processing-capable hardware. OpenCV provides productivity for professionals in achieving the best results in the field of vision processing. Initially, image processing was only possible in research labs due to its rare availability on open source platforms and dedicated hardware. Nowadays, students, researchers and professionals can easily use the fully developed and growing library known as OpenCV. OpenCV, at the time of writing this book, is of version 3.0 Beta, with approximately 9 million downloads.

Camera interfacing with the RasPi

The RasPi has the Broadcom multimedia processor BCM2835, which has an integrated Videocore-IV graphics engine. There are multiple ways to interface the camera with the RasPi connectors. As described in *Chapter 1, Meeting Your Buddy – the Raspberry Pi,* (refer to the diagram showing the RasPi's port description), the RasPi has a dedicated CSI port for connecting the camera module. This port can be located near the Ethernet controller (in the case of model B) or near the 4-pole audio connector (in the case of model B+). The **Camera Serial Interface (CSI)** port is dedicated to camera interfacing in mobile devices and advanced peripheral controllers. We have two options to interface the cameras with our RasPi. We can either interface the webcam via USB, or choose the camera module from the Raspberry Pi organization known as the **Raspberry Pi camera module**.

The RasPi camera modules

There are two options when buying the dedicated camera modules to be interfaced with RasPi. The camera comes with the half-foot long flex cable to connect it with the RasPi CSI connector. The RasPi camera modules are available in two options:

- **Raspberry Pi camera module**: This camera module employs a 5-megapixel camera with the capability of full HD video recording at 30 FPS. It does the job well during the daytime with better light.

- **Raspberry Pi camera module black NoIR**: **NoIR** stands for **No Infrared**. This camera module doesn't have an internal infrared filter, which gives the camera the ability to see in the dark. You can deploy this camera as a CCTV in your garden house to keep a watch on your pet at night. These camera modules perform even better for wildlife photographers by triggering and capturing the image or video whenever motion is detected. This camera has the ability to capture images at night because it deploys infrared LEDs.

These cameras have a price tag of $25 in online stores. To enable the RasPi camera module, enter this command and enable the camera from the configuration window:

```
sudo raspi-config
```

Select the **Enable Camera** option in the configuration settings using the keyboard. Test the camera by capturing the images using the following command:

```
Sudo raspistill -o FirstimageThroughRasPi.jpg
```

Use the `man raspistill` command to know about more options available with the `raspistill` command. You can even record videos (in H264 format) using the following command:

```
raspivid -o FirstVideoThroughRasPi.h264 -t 5000
```

This command will record a video for 5 seconds, and will automatically save the file when the recording finishes.

 The RasPi camera modules are very sensitive to electrostatic charges transferred by our body. RasPi camera module failures are very common in the Raspberry Pi community. Wear electrostatic straps or taps on your hand before touching the camera. The ribbon supplied with the RasPi camera module is very delicate. Therefore, special care should be taken while using this camera module.

USB webcam

It can be cheaper if you interface your desktop computer's webcam by compromising performance. Not all webcams are supported on the RasPi. You can check out the list of webcams and cameras supported by Linux at http://www.ideasonboard.org/uvc/#devices.

There is a trade-off between speed, cost, and versatility by using a USB webcam compared to the Raspberry Pi camera module. If you choose a USB camera, you will get degraded performance, but you will be able to interface a USB webcam with any of your single-board embedded computers in future. USB cameras are definitely cost-effective. Added to that, a strong USB cable and static-proof casings provide a longer life compared to the Raspberry Pi camera module.

The upcoming project in this chapter is verified using the Logitech C270 USB web camera.

Live streaming using a network camera

Have you ever wondered whether the RasPi can be used for live streaming purposes? We are going to stream videos on the RasPi on the localhost network using RasPi as a network camera. There are multiple ways to make use of the RasPi for live streaming:

- By fixing a camera interfaced with the RasPi being in the same network, you can SSH through the connection and open the camera port through the command-line interface

- Use open source software motion to enable live streaming at the local web browser

We will be using motion to get access to the RasPi camera. Motion has a built-in HTTP server that enables the camera view from the web browser. Motion can record videos in MPEG format and capture images in JPEG format. These images and videos can be stored wherever you want, and it does support databases such as MySQL. Enter these commands to install motion on the RasPi:

```
sudo apt-get update
sudo apt-get upgrade
sudo apt-get -y install bison
sudo apt-get -y install motion
sudo apt-get -y install v4l-utils
```

It is required to set the configuration file before starting the project, which can be called IP-camera (network camera):

```
sudonano /etc/motion/motion.conf
```

Setting the configuration file could be a tedious job, as the file is too long and there are many parameters to set. You can search parameters in the file using *Ctrl + W* in nano editor. Set the parameters represented as follows:

```
netcam_url value http://169.254.0.2:8081
webcam_localhost off
control_localhost off
control_port 8080
framerate 2
control_html_output on
post_capture 5
daemon on
```

 The RasPi's IP address can be found using the `ifconfig eth0` command. In response message, it must have been noted in the second line as `inet addr:169.254.0.2`.

We have informed the IP address of the RasPi module and provided a port address for motion, where the stream will be diverted. By setting the webcam and control localhost as off, the RasPi will be able to give access to other computers in the network on ports 8080 and 8081. Setting the frame rate at more than 6 or 7 FPS will be computationally heavy for RasPi to handle, and it may crash during runtime. Setting the frame rate at 2 FPS will provide stable performance. Setting the control HTML output will provide the functionality of that HTML server's built-in motion API. Whenever motion is detected, the `post_capture` setting will let the RasPi know the number of frames to be captured. To start the motion service when the RasPi boots, start the daemon by setting it as on.

After making these changes in the configuration file, press *Ctrl + X*, then *Y*, and then the *Enter* key to save and exit the configuration file. There is a need to change the daemon file of the motion to keep the motion service running. Enter this command to set the daemon for motion:

```
sudo nano /etc/default/motion
```

Edit `start_motion_daemon=yes`. Save and exit the file. After all of these settings are completed, create a folder and run the following command to start the motion service:

```
mkdir motion-camera

cd motion-camera

sudo service motion start
```

Now, on the PC desktop, open a web browser and type the IP address and port number as saved in the motion configuration (`motion.conf`) file:

```
http://169.254.0.2:8081
```

You will be able to see live streaming from the RasPi camera. Congratulations! We have made our own network camera. This camera can be accessed from any computer in the network. All you have to do is connect the camera in your backyard through an Ethernet cable. You can directly access it from your living room. You can even try it from VLC media player. In VLC, go to **Media** and then go to **Open Network Stream**. Enter the same IP address, followed by the port address.

Porting OpenCV

Porting OpenCV on RasPi is the lengthiest task compared to installing other libraries. Installing the suitable and companion libraries is necessary for effective running of the image processing algorithms. It has been observed in the community that not many have succeeded in porting the OpenCV properly. We will go through a step-by-step guide that will give you an up-and-running OpenCV at the first time. While installing some of the library, the order of the installation doesn't matter, but it is suggested that the flow given here will give you the exact results you are looking for.

A quick checklist is as follows:

- An Internet connection shared with RasPi through a Wi-Fi adapter or a PC.
- Use PuTTY to log in to the RasPi using an Ethernet connection from a PC.
- Enough space (up to 4 GB) on the RasPi SD card. Use the `df -h` command to check out the free space on the SD Card.
- Interface the camera when needed. A USB camera needs enough current to operate. Connect a 2A adapter to the RasPi if needed.

Let's start! Log in to your RasPi on the SSH connection over the Ethernet. From now on, we will be using PuTTY to enter all of these commands in the RasPi unless otherwise specified.

This installation may take up to 12 hours to build the OpenCV on Raspbian. At any stage, resumption of installation after interruption can be done without reinstalling the previous libraries.

Ensure that you have enabled SSH communication on the RasPi. You can enable it using this command, stating the IP address of your RasPi:

```
ssh -X pi@169.254.0.2
```

Type Yes and then press *Enter*, and SSH will be enabled on your Pi. You can continue in the same session. You can use this command to enable SSH communication over the Ethernet:

```
sudo raspi-config
```

Then enter **Advanced Options** and find the option mentioned as SSH. Press *Enter* and enable the option. In the same menu, there is an option called **Camera**; this should be enabled as well. Press *Esc* to exit the configuration menu.

Now it is time to update the RasPi to provide information on the latest packages, versions, and dependencies. All the repositories will get information about their latest packages and information for resynchronizing:

```
sudo apt-get update
```

Next, upgrade will fetch new versions of packages according to the list provided in the update list:

```
sudo apt-get upgrade
```

To get the packages, you can install the synaptic package manager. It allows you to install packages using the GUI upfront instead of entering the sudo apt-get command. It will be handy to install the missing packages once OpenCV is installed:

```
sudo apt-get -y install synaptic
```

Note that we will still be using the command-line interface to install the packages, as it is the better way to install entire libraries. You can use synaptic but prefer installing small packages and managing repositories. It will be better to manage the installation in the command-line interface.

We have to install the scientific and mathematical libraries of Python to process the images and to get the additional functionalities. Numpy and Scipy are the mathematical and scientific libraries used to calculate *n*-dimensional arrays, Fourier transforms, random numbers, and linear algebra calculations. Nose is the library used to test Python code in an easier and faster way. Next, Pandas is a high-performance data analysis tool for Python. A role of IPython is to provide functionalities in an interactive way to see the results of the code and commands, with interactive visualization on the GUI. Matplotlib is used to plot two-dimensional graphs, which provides good-quality graph images. These can be used in studies and publications. All the libraries can be added in a single line and can be installed in one go. Finally, SymPy provides the functionality of symbolic mathematics. Its purpose is to provide functions in the computer algebra system, while keeping the Python code as simple as possible, and extensible. All of these libraries can be used while coding with OpenCV using the Python library. Enter this text in a single line to execute in the command-line interface:

```
sudo apt-get -y install python-numpy python-scipy python-nose python-pandas python-matplotlib ipython-notebook python-sympy
```

GtkGLExt needs to be installed to provide support for OpenGL rendering. OpenGL is an API that is typically used by GPUs to achieve 3D and 2D graphics. We will be enabling it while building OpenCV. Use the following command to install GtkGLExt:

```
sudo apt-get -y install libgtkglext1-dev
```

To build the development environment in the RasPi and make all the essential libraries up to date and compliable in the Linux OS, use CMake to configure the packages and to manage the build process. Enter the following command:

```
sudo apt-get -y install build-essential cmake pkg-config
```

Qt is widely used as a cross-platform application framework to develop applications, which can be developed and run on various hardware and software platforms with minimal or no changes in the code base, keeping the native application as it is. Enter this text in a single line to execute and install the necessary packages of Qt in the command-line interface:

```
sudo apt-get -y install qtcreator qt4-dev-tools libqt4-dev libqt4-core libqt4-gui v4l-utils
```

If you are enthusiastic enough to code in Java, you should install the Ant library. OpenCV has inbuilt Java support. To enable the functionalities of compiling, building the `.jar` files, and utilizing it efficiently, type the following command:

```
sudo apt-get -y install default-jdk ant
```

We are now ready with all the necessary packages to start downloading the OpenCV library. You can download the latest OpenCV library 3.0.0, which is in beta version at the time of writing this book. I prefer to install the latest stable version (2.4.10). Later on, we can upgrade OpenCV. Enter this command to download the library:

```
wget http://sourceforge.net/projects/opencvlibrary/files/opencv-
unix/2.4.10/opencv-2.4.10.zip
```

It may take a while to download a file from the official source. Once it is downloaded successfully, you can unzip the file using the following command:

```
unzip opencv-2.4.10.zip
```

Alternatively, if the file downloaded is the tarball, then execute this command:

```
tar xzvf opencv-2.4.10.tar.gz
```

Executing this command will show you the entire list of files transferred to the folder. After unzipping the files, they are extracted to the opencv-2.4.10 folder. Create a build folder at the same location. Enter these commands sequentially:

```
cd opencv-2.4.10
mkdir build
cd build
```

Next, we are going to use the CMake command to build the configuration type for release. Look for the output of CMake and check which packages are installed and which are not. You can install those packages manually. By looking at the CMake output, you fill find YES and TRUE in front of all the libraries and packages. If not, do not panic! Just list down those names and install these dependencies using the synaptic package manager. It is very important to put .. at the end of the command to successfully create a make file:

```
cmake -D CMAKE_BUILD_TYPE=RELEASE -D WITH_OPENGL=ON -D INSTALL_C_
EXAMPLES=ON -D INSTALL_PYTHON_EXAMPLES=ON -D WITH_QT=ON -D CMAKE_INSTALL_
PREFIX=/usr/local -D WITH_TBB=ON -D WITH_V4L=ON -D BUILD_NEW_PYTHON_
SUPPORT=ON -D BUILD_EXAMPLES=ON ..
```

 To use the synaptic package manager, use Xming, open the desktop by typing lxsession in PuTTY, and select synaptic from the menu in GUI. Then type the library name in the search bar and install the additional libraries or dependencies.

The next step is to compile the entire library. As we have seen, the OpenCV library has many components.

 Before executing the `make` command, open the case of the RasPi and put it in a place where it gets enough air to cool down. It is normal that the RasPi gets heated during this operation.

Using the previous command, `cmake`, we generated an intrinsic make file, which will be used in the environment of our choice. Due to dependencies between the components in the library, the `make` command will be useful for execution:

```
make
```

Now, if it is evening when you are working on this installation, grab some popcorn, watch your favorite TV show, and take a beautiful sleep. You will be able to see the results by the next morning! As mentioned earlier, this compilation on the RasPi takes almost 10 to 14 hours to complete . Do not disturb the RasPi while it is compiling the libraries. Do not worry if the compilation halts or sticks somewhere in the middle, or even if your RasPi shuts down accidently. Start the session again, go to the same build folder, and enter the `make` command again. It will resume from where it was stuck. RasPi 2 model B performs exceptionally well here. This compilation takes 3 to 4 hours to finish the `make` process on RasPi 2 platform.

Once this lengthiest task is done, we are ready to install the compiled library on the RasPi. This task will not take as much time as it took in the previous step:

```
sudo make install
```

It's time to set the configuration file for OpenCV to ldconfig in `/etc`. The ldconfig creates the essential links to the latest shared libraries found in the directories stated in the command line, in the `/etc/ld.so.conf` file and in the `/lib` and `/usr/lib` paths. Basically, it tells your RasPi's OS that we have installed the library. Therefore, we will create a separate file for the OpenCV configuration:

```
sudo nano /etc/ld.so.conf.d/opencv.conf
```

Enter the following lines in the configuration file we just opened:

```
/usr/local/lib
```

Then exit the file using *Ctrl + X* and then pressing *Y* followed by *Enter*.

Enter the `sudo ldconfig` command to effect the changes made in the file. In the interactive shell source (which brings the terminal emulator) file, we will paste the code at the end of the file. The file is quite long, so use the keyboard to scroll down:

```
sudo nano /etc/bash.bashrc
```

Enter these lines at the end of file, save, and exit. Note that the `export` command should be entered in a new line:

```
PKG_CONFIG_PATH=$PKG_CONFIG_PATH:/usr/local/lib/pkgconfig
export PKG_CONFIG_PATH
```

Editing this file gives us the freedom to compile the OpenCV code in any directory. It is similar to putting the environment variable for your library in the Linux operating system.

Let's check whether this entire setup works or not.

Testing

It was so tiring to install the OpenCV, but the next experiment will fill you with joy, and you will experience that the RasPi is finally able to see you and detect your face! It will be difficult to explain OpenCV in this section. You will directly execute the sample program written in Python to detect your face. But before we start executing, we will test our camera to check whether it is working or not. If you have interfaced your camera on one of the USB ports, type this command to check the camera name in the USB list:

```
lsusb
```

This will list all the USB devices connected on the port. Now install the software to open the camera port and take the images:

```
sudo apt-get -y install guvcview
```

To capture the images, you have to enter the desktop mode of RasPi, as `guvcview` is a GUI-based tool. Open the Xming server in your PC and enter the following command in PuTTY that will open the real-time desktop of the RasPi in your PC:

```
lxsession
```

When coding OpenCV, it is recommended to open all the sessions in Xming using PuTTY. Xming allows you to open the windows to show the images captured during the execution of the OpenCV code.

This command will open the RasPi desktop environment and then the LxTerminal from the desktop. Now open the installed guvcview GUI by typing the following command:

```
guvcview
```

This open source software lets you capture images and record videos, along with sounds (well, we don't have microphone on the RasPi, but USB webcams do have integrated microphones). After testing the camera interfaced with the RasPi, move to the OpenCV installation folder using the following command:

```
cd /home/pi/opencv-2.4.10/samples/python
ls
```

You will see a number of programs listed in the folder, as follows:

camera.py	dft.py	houghlines.py	morphology.py
camshift.py	distrans.py	inpaint.py	motempl.py
chessboard.py	dmtx.py	kalman.py	numpy_array.py
contours.py	drawing.py	kmeans.py	numpy_warhol.py
convexhull.py	edge.py	laplace.py	peopledetect.py
cv20squares.py	facedetect.py	lkdemo.py	pyramid_segmentation.py
cvutils.py	fback.py	logpolar.py	squares.py
delaunay.py	ffilldemo.py	minarea.py	watershed.py
demhist.py	fitellipse.py	minidemo.py	

Play around with any of these programs, and do not forget to test facedetect. py and camera.py. It is also interesting to play with contours.py, which brings an interactive GUI with sliding scrollbars for setting the contours.

Adjust the camera facing it towards you. Execute the face detection program by simply writing this command:

```
python facedetect.py
```

Yes, it is slow! The detection time is about 5 to 7 seconds in RasPi 1 models. You can try the same experiment with the Raspberry Pi camera module, and a drastic change in the performance of RasPi will be observed, as the detection time falls to 2 seconds or less. With RasPi 2 model B, it's even lesser. Anyway, you can edit and reduce the frame size to make it real-time after editing the program.

Create a motion detector

We have already set up a piece of software called motion. It detects and captures motion, but that's not fun at all compared to building our own code to capture motion. We will create code that will detect motion and take a decision according to the extent of motion. We will trigger an LED to indicate the decision taken by the RasPi. This code is going to be a bit long, so the explanation is given in comments within the code. However, important information is provided just after the end of the code:

```
#include <iostream>
#include <cv.h>
#include <highgui.h>
#include <stdio.h>
#include <cctype>
#include <iterator>
#include <unistd.h>
#include <wiringPi.h>
#include "opencv2/highgui/highgui.hpp"
#include <time.h>

using namespace cv;
using namespace std;

int main()
{
/*----Section 1 ----------------Declarations----------------*/
wiringPiSetup();  //To toggle LED, Essential
pinMode(0, OUTPUT); //BCM_GPIO pin 17
CvMemStorage* g_storage = NULL; //clearing the memory
int i,count=0; //Declaring the functions
CvCapture* imagecapture = cvCaptureFromCAM(CV_CAP_ANY);
```

```
while(1)
   {
/*----Section 2------current frame capture started--------------*/
     IplImage* current = cvQueryFrame(imagecapture);
     cvSaveImage("/home/pi/project/current.jpg",current);
     //cvNamedWindow("LIVE VIDEO TRACKING", 1);
     //For debugging only
     //cvShowImage("LIVE VIDEO TRACKING", current);
     cvWaitKey(150);
     IplImage* img1= cvLoadImage("/home/pi/current.jpg");

/*----Section 3------Reference frame capture started------------*/
     IplImage* reference=cvQueryFrame(imagecapture);
     cvSaveImage("/home/pi/Reference.jpg",reference);
     IplImage* img2=cvLoadImage("/home/pi/Reference.jpg");
     IplImage* img3=cvLoadImage("/home/pi/Reference.jpg");
     //to give the same size as img1 and img2
     cvAbsDiff(img1,img2,img3);

/*----Section 4----------THRESHOLD, ERODE, DILATE--------------*/
     IplImage* result = cvCreateImage(cvGetSize(img3),IPL_DEPTH_8U,1);
     //Creating new image as the same size of img3
     cvCvtColor(img3,result,CV_BGR2GRAY);    //Convert to Grayscale
     cvThreshold(result,result,10,255,CV_THRESH_BINARY); //Convert to
binary image and update in same variable
     //cvShowImage("THRESDOLEDE IMAGE",result); //For Debugging
     cvErode(result,result,0,2);         //Erosion of Image
     cvDilate(result,result,0,2);   //Dilation of Image

/*----Section 4-----------AVERAGE AND AREA--------------------*/
     CvScalar avg = cvAvg(result); //Counting and taking average of
pixels
     int area;
     area = (int)avg.val[0];
     printf("Area of object is : %d \n", area);
     char *date;
     char buffer[1000];

/*----Section 5-----------MAKE DECISION----------------------*/
     if (area >= 25 && area<=150) //Typical Human Area within 3 meters
        {
         digitalWrite (0, 1);
         printf("Intruder Detected");
         cvNamedWindow("Intruder",0);
         cvResizeWindow("Intruder",320,240);
```

```
        cvShowImage ("Intruder",img1);

/*----Section 6----Saving Files, Releasing Memory-------------*/
        time_t timer;
        timer=time (NULL);
        date = asctime(localtime(&timer));
        sprintf(buffer, "/home/pi/intruder_%s.jpg",date);
        //Appending Image Name with Date
        cvSaveImage(buffer,current);//Saving Image with Name-Date
        delay (150);
        }
    else
        {
        digitalWrite (0, 0);
        cvDestroyWindow("Intruder");   //Closing the Window
        delay (150);
        }
    cvReleaseImage(&img1);    //Releasing the memory used
    cvReleaseImage(&img2);
    cvReleaseImage(&img3);
    cvReleaseImage(&result);
    if (cvWaitKey(100)== 27)
    break;
    }
return 0;
}
/*----End of the Code----------------------------------------*/
```

To detect motion, we will use the simple method of image subtraction. We will capture a reference image at a particular interval, and subtract the instantaneous frames coming from the camera. The difference between these images will be sent to a threshold function to observe the change. This change will then be input to the function to calculate the absolute area of observed motion. If the area is in the range of 25 to 150 units (neither too large nor too small; this is experimented using different values), then the suitable action will be taken. These values are optimum for use in detecting human motion. So, if there is any movement in the area of interest, you can detect it as well as capture the image of the intruder. Let's understand the code to get an idea of the entire methodology. We will be using the `wiringPi` library to trigger the LED whenever motion is detected.

 OpenCV functions are case-sensitive; for example, the `cvSaveImage()` function will work only when the capitalization of characters in the word is proper.

The code can now be saved as `motion.cpp` in the home folder or a place of your choice. The names of the functions used in this code are self-explanatory enough, though some of the functions and variables are listed here with their explanations. Take a look at the description of following commands used in the code:

IplImage*

Whenever the `cvLoadImage` or `cvQueryFrame` function is called, the pointer to an allocated image data structure is returned. This pointer will be used as a reference to an image. Basically, this shows the memory location of the first pixel of an image.

cvCreateImage(cvGetSize(img3),IPL_DEPTH_8U,1)

This creates an image pointer and allocates the image data from the pointing address to that image. There are a total of three fields, in which the first is the size of an image. In the first field, we have called a function that gets the size of an already declared or saved image. The second field in the function defines the bit depth of an image, just as `IPL_DEPTH_8U` defines the unsigned 8-bit integer depth for a single pixel. For higher quality images, you can use `IPL_DEPTH_64F`, which defines the double-precision floating point. Consider an 8-bit unsigned integer while working on the RasPi, as processing can be done smoothly. The third field defines the number of channels in the image.

cvCaptureFromCAM(CV_CAP_ANY)

This function opens the camera ports and takes images from the interfaced camera modules. Entering `CV_CAP_ANY` allows OpenCV to choose the cameras interfaced. The RasPi will choose the camera according to the priority of the interfaces. This function works in the same way as `CvCapture* image = cvCreateCameraCapture(CV_CAP_ANY)`. If multiple camera modules are interfaced and you want to choose the camera before capturing, you can use `cvCreateCameraCapture(-1)` to open a window that will allow you to select the camera modules interfaced.

cvAbsDiff(img1,img2,img3)

This function calculates the absolute difference between two image arrays and stores it in the destination image. Here, the result will be stored in `img3` by performing the `img3 = img1 - img2` operation. This will be helpful to us for detecting the change in the current image by subtracting it from the reference image.

cvCvtColor(img3,result,CV_BGR2GRAY)

The function can be useful for converting the color space of the `img3` image and to store the resulting image. The third field in the function is the color space conversion code. This code will convert the RGB (red, green, and blue) image into a grayscale image.

 Note that the default color format in OpenCV is often mentioned as RGB, but it is actually BGR (the bytes are swapped).

`cvThreshold(result,result,10,255,CV_THRESH_BINARY)`

Mostly, the threshold function is used to convert a grayscale image into a binary image. If the pixel value is less than the threshold value, then replace that pixel with `0` (black); otherwise, use `1` (white). There are a total of five fields in the function. The source and the destination are the first two fields. The third is the threshold value, which can be tweaked with different experiments. If `CV_THRESH_BINARY` is used, then the maximum value to be used is `255` (white), which is defined by the fourth field. There are multiple options available in this, but this function helps us convert it into a binary image. The function for the binary image threshold can be defined as `result(x, y) = 255` if `img(x, y) > 10` otherwise `result(x, y) = 0`.

`cvErode(result,result,0,2)`

In many applications, an erosion function is performed to remove salt noise from the image. Due to the threshold operation on the image, there could be noise in the image pixels, which can create a false result while taking the average value of an image. In our application, the erosion function deletes the boundaries of the white object. Therefore, the small white dots will be completely removed from the image after performing two iterations. There are four fields, with source and destination as the first two fields. The third field defines the structuring element, which will be zero in our case. The fourth element states the number of erosion iterations to be performed on an image, like this:

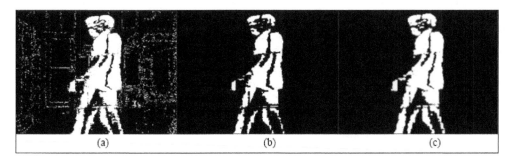

(a)　　　　　(b)　　　　　(c)

The preceding image has three sections. Image **(a)** is the result after the execution of the `cvAbsDiff` and `cvThreshold` functions. The movement of the intruder can be seen in the image. The white salt noise is clearly visible after the threshold function. Image **(b)** is the result after performing the `cvErode` function. This removes the salt noise by erasing the boundaries of white objects or pixels in the image. Finally, image **(c)** is the result after the `cvDilate` function, which is described next:

```
cvDilate(result,result,0,2)
```

A dilation function is the reverse of an erosion function. After removing the salt noise, the dilation function improves the group of white pixels at the edges of an image.

The code flow can be understood in this way: section 1 in the code defines the variables and captures an image called `imagecapture`, which will be used as the current image captured by a camera in section 2. Next, section 2 starts with an infinite loop, and we are saving the current image and then loading it back into a variable called `img1`. The commented code can be uncommented when debugging to track the correctly captured images by triggering the window-opening function. In section 3, the reference image is saved and loaded back into the `img2` variable. The absolute difference between the reference and the current image will be taken, and will be stored in `img3`. This can create some confusion; before taking the absolute difference, why is the reference image being loaded into `img3`? The reason behind this is that `img3` needs the same resolution as `img1` and `img2`. Otherwise, OpenCV will show an error of mismatch and that the subtraction could not be performed. Section 4 is dedicated for image processing functions, such as converting an image to grayscale and performing a threshold operation on an image to make the image binary. Section 5 takes the decision as to whether to trigger the GPIO or not. Section 6 is all about saving the files and releasing the memory. Keeping the delay between the operations is very important because the RasPi and OpenCV take their own time to perform operations and save the files. That's all! We are ready to compile the code now.

Preparing shell to compile OpenCV and wiringPi

After preparing the code, it could be difficult for us to debug and compile it unless we prepare a generic shell script. The use case of this shell file is to compile all the available C and C++ files in the same folder. Whether the code is in C or C++, if the code contains the OpenCV and wiringPi libraries, it can be compiled to generate an output file with the same name as the code filename.

For example, suppose you are working in a folder that contains the code files named
project0.cpp and project1.c. If you execute the shell file being in the same folder,
it will start compiling all of the available C code by generating the executable files as
project0 and project1, which can be later run by ./project0. Let's experiment
with the shell code. Go to the folder where you wrote the motion detection code
(motion.cpp), open nano editor with the build.sh filename, and type this code:

```
#!/bin/sh
if [ $# -gt 0 ] ; then
    base=`basename $1 .c`
    echo "compiling $base"
    gcc -ggdb `pkg-configopencv --cflags --libs` $base.c -o $base
else
    for i in *.c; do
        echo "compiling $i"
        gcc -ggdb `pkg-config --cflags opencv` -o `basename $i .c`
$i  -lwiringPi `pkg-config --libs opencv`;
    done
    for i in *.cpp; do
        echo "compiling $i"
        g++ -ggdb `pkg-config --cflags opencv` -o `basename $i .cpp`
$i -lwiringPi `pkg-config --libs opencv`;
    done
fi
```

Save this shell file by exiting nano editor. The shell code written here is derived from
the OpenCV samples folder and edited according to our needs. The wiringPi library
is mentioned in bold characters. The if-else loop is classified as a one-file and
multiple-files compilation. This code will take each C and C++ file one by one and
will save the executable. Now give access to execute this file:

chmod +x build.sh

Execute the shell code by typing ./build.sh. It will start compiling all the .cpp and
.c files in the folder. Other files in the folder will remain unaffected. The executable
output file will be saved as motion. Execute the file by typing ./motion to see the
output. Remember to execute this file by opening LxTerminal in the Xming server to
see the window.

Amazing projects for you

With some knowledge on OpenCV, you can go ahead and read the manuals of OpenCV to gain more knowledge on how functions and arguments are passed. Initially, it can be somewhat difficult for you to start OpenCV using image pointers. There is an immense amount of help available at http://www.opencv.org. While taking help from the Internet and the code available from the official website, check for the OpenCV language (platform) first. It has been observed that when C code merges with C++ code, it doesn't work even after banging your head for hours. It is advised to keep *The OpenCV Reference Manual* (Current Release 2.4.9.0) with you while coding OpenCV.

It would be quite a good challenge for you to develop a project that fulfils the following requirements:

- Capture live images using the interfaced camera, split the image into segments of the same size (say, four vertical segments), and compare it with the reference image. There will be a clear difference in one of the segments where most of the motion is observed. Detect that area in the image and trigger the GPIO accordingly.

- Further, control a servomotor or stepper motor to rotate the camera according to the movement in front of it. Finally, you will be able to develop a camera that follows you wherever you move.

- Another application of this is to save energy. Split the image into coordinative segments and detect where people are sitting in the room by setting the camera at the heighted position. Turn off the light where zero motion is detected. Interesting, isn't it?

- You can develop OpenCV code to track a particular colored object. The code can draw the lines on the screen while this colored object is moved in front of a camera. For example, when we show and move a red ball in front of the camera lens, it starts drawing lines as per the red color's movements detected by camera.

- Create a gesture/pattern-recognizable computer module that takes commands as per the color rings worn on your fingers.

- Build a ball-following robot that can follow a ball of a particular color and detect the near and far effects of the ball.

Use a separate motor and relay driver boards to control the motors and electrical appliances, instead of directly driving them from GPIO. Be aware of electrical shocks while working with relay-based appliance control.

After building these projects, you can show them to your friends and be the coolest person among them. It can be really very interesting to build these projects, can't it? Then build it!

Summary

This chapter was totally dedicated to development in image processing using the most advanced OpenCV library. We started with an overview of image processing and its applications, followed by an introduction to OpenCV. By now, you already know that the OpenCV library is so vast that even four books like this would not be enough to provide a detailed description of the entire library. You then understood the camera interface on the RasPi module, and we chose the USB camera to go with. Next, you experimented live streaming using the RasPi as a network camera module (IP camera) to keep track of your backyard directly from the lounge.

We started porting the OpenCV library, which took almost half a day to be up and running. Then, we verified the library using the camera interface. Next, we started building the project in the C language to detect human motion by calculating a motion-affected area. A useful shell file was prepared to compile the OpenCV code along with the wiringPi interface. A task was given to you to build the project, prepare it, and send it for the suggestions on the code.

With the end of this chapter, this book is pretty much complete. Looking at the current trends, the Raspberry Pi world will keep growing. With support from thousands of people in community, the RasPi has achieved the highest position in the market of single-board embedded computers.

With the launch of RasPi 2 model B, a new era in computing has begun. OpenCV libraries and image processing now have a better performance on Raspberry Pi module 2. RasPi users will be able to enjoy totally free Windows 10 on this module. You can give it a try once Windows 10 is available. Kudos to the Raspberry Pi organization and their engineers for pouring hours into the development of these modules!

As you have experienced, the RasPi has infinite use cases. You have to explore different possibilities with it. You can hook up a camera with the simplest LED to display indications. You can interface a 7-inch touch LCD to make it a personalized Android tablet or a full-HD LED TV to watch videos. With the freedom of Linux, starting from wiringPi and Python, you can use the awesome libraries such as Node.js, Apache, C++ REST, Qt, Mathematica (or the Wolfram Language), OpenCV, and many more. The RasPi can be used to make a project such as toggling the LED, or a more complex project such as making your own quadcopter with stability control algorithms.

I hope that you will make most out of it. Be curious and keep learning.

Shine on! All the best!

Shopping List

Basic requirements

- A Raspberry Pi (Raspberry Pi 1 model B, Raspberry Pi 1 model B+, Raspberry Pi 1 model A+, or Raspberry Pi 2 model B (Purchase any one))

- Micro SD card with an SD adapter with storage capacity of 8 GB or more

- Personal computer with a Windows, Mac OS X, or Linux environment

- Ethernet cable (RJ45)

- A 5V 1A or 5V 2A power adapter with micro-USB connector

- HDMI or RCA cable

- Breadboard

- Multimeter

- Wire stripper

Sensors

- HC-SR04 ultrasonic sensor (Quantity: 1)

- DHT11 temperature-humidity sensor (Quantity: 1)

- LDR or CdS photocell or photoresistor (Quantity: 1)

- TMP35, LM35, or TMP36 temperature sensor (Quantity: 1)

Integrated chips

- Analog-to-digital convertor (Quantity: 1): MCP3008 dual-in-line package or MCP3004 dual-in-line package

Components

- 1/4 Watts through hole resistors (Quantity: 5 each): 1KΩ, 2KΩ, 270Ω, 330Ω, 470Ω, 4.7KΩ, and 10KΩ
- Through-hole electrolytic capacitor (Quantity: 5 each): 1μF-16V
- LEDs (Quantity: 5 each): 2 mm, 3mm, or 5mm Red/Green/White/Yellow LED

Others

- Single stranded wire (Quantity: 1 meter)
- Female-to-male jumper wires (Quantity: 15) and female-to-female jumper wires (Quantity: 15)
- Bergstik connectors (Quantity: 2 each): Dual-row male, 2.54 mm pitch single-row male, 2.54mm pitch
- Female-to-female GPIO ribbon cable for RasPi 1 model B (26 pin), B+ (40 pin), or RasPi 2 model B (40 pin) (Quantity: 1)
- General purpose circuit board: dual sided and solderable
- Pencil type soldering iron (30-50 Watts)
- Soldering wire with flux (Quantity: 50 gms)
- Camera (Quantity: 1): Logitech C270 USB webcam or Raspberry Pi camera

Index

Thank you for buying
Raspberry Pi Sensors

About Packt Publishing

Packt, pronounced 'packed', published its first book, *Mastering phpMyAdmin for Effective MySQL Management*, in April 2004, and subsequently continued to specialize in publishing highly focused books on specific technologies and solutions.

Our books and publications share the experiences of your fellow IT professionals in adapting and customizing today's systems, applications, and frameworks. Our solution-based books give you the knowledge and power to customize the software and technologies you're using to get the job done. Packt books are more specific and less general than the IT books you have seen in the past. Our unique business model allows us to bring you more focused information, giving you more of what you need to know, and less of what you don't.

Packt is a modern yet unique publishing company that focuses on producing quality, cutting-edge books for communities of developers, administrators, and newbies alike. For more information, please visit our website at www.packtpub.com.

About Packt Open Source

In 2010, Packt launched two new brands, Packt Open Source and Packt Enterprise, in order to continue its focus on specialization. This book is part of the Packt Open Source brand, home to books published on software built around open source licenses, and offering information to anybody from advanced developers to budding web designers. The Open Source brand also runs Packt's Open Source Royalty Scheme, by which Packt gives a royalty to each open source project about whose software a book is sold.

Writing for Packt

We welcome all inquiries from people who are interested in authoring. Book proposals should be sent to author@packtpub.com. If your book idea is still at an early stage and you would like to discuss it first before writing a formal book proposal, then please contact us; one of our commissioning editors will get in touch with you.

We're not just looking for published authors; if you have strong technical skills but no writing experience, our experienced editors can help you develop a writing career, or simply get some additional reward for your expertise.

Raspberry Pi Cookbook for Python Programmers

ISBN: 978-1-84969-662-3   Paperback: 402 pages

Over 50 easy-to-comprehend tailor-made recipes to get the most out of the Raspberry Pi and unleash its huge potential using Python

1. Install your first operating system, share files over the network, and run programs remotely.

2. Unleash the hidden potential of the Raspberry Pi's powerful Video Core IV graphics processor with your own hardware accelerated 3D graphics.

3. Discover how to create your own electronic circuits to interact with the Raspberry Pi.

Raspberry Pi Robotic Projects

ISBN: 978-1-84969-432-2   Paperback: 278 pages

Create amazing robotic projects on a shoestring budget

1. Make your projects talk and understand speech with Raspberry Pi.

2. Use standard webcam to make your projects see and enhance vision capabilities.

3. Full of simple, easy-to-understand instructions to bring your Raspberry Pi online for developing robotics projects.

Please check **www.PacktPub.com** for information on our titles

Raspberry Pi Server Essentials

ISBN: 978-1-78328-469-6   Paperback: 116 pages

Transform your Raspberry Pi into a server for hosting websites, games, or even your Bitcoin network

1. Unlock the various possibilities of using Raspberry Pi as a server.

2. Configure a media center for your home or sharing with friends.

3. Connect to the Bitcoin network and manage your wallet.

Raspberry Pi for Secret Agents

ISBN: 978-1-84969-578-7   Paperback: 152 pages

Turn your Raspberry Pi into your very own secret agent toolbox with this set of exciting projects!

1. Detect an intruder on camera and set off an alarm.

2. Listen in or record conversations from a distance.

3. Find out what the other computers on your network are up to.

4. Unleash your Raspberry Pi on the world.

Please check **www.PacktPub.com** for information on our titles